INDUSTRIAL AUTOMATION WITH SCADA

Concepts, Communications and Security

K S MANOJ

INDIA · SINGAPORE · MALAYSIA

Notion Press

Old No. 38, New No. 6
McNichols Road, Chetpet
Chennai - 600 031

First Published by Notion Press 2019
Copyright © K S Manoj 2019
All Rights Reserved.

ISBN 978-1-68466-828-1

I owe this book to my dear sir... Prof. S. Suryadas, who introduced me to the exciting world of Electronics and Communication.

CONTENTS

PREFACE

AUTHOR'S NOTE

It is a fact that SCADA has become the synonym of automation in industry. In order to implement these technological requirements, automation engineers who are presently engaged in the implementation of SCADA or Internet of Things, and the Instrumentation engineering students who are intended to be a part of the evolving IoT technology, a minimum and essential understanding of distributed SCADA system, guided and unguided communication technologies, Network security, SCADA protocols and interoperability, etc. are very much essential. Without which the Instrumentation engineers and the ICT professionals move forward parallel and never converge to attain the desired goal. This book is mainly intended to equip the Instrumentation Engineers with the basic concepts required to design and develop Distributed Control Systems using SCADA architecture. Further many SCADA systems are employed in critical infrastructure of a Nation, the communication technology requirement is not just communication rather, secureCommunication (sCommunication). Hence emphasis has been given to describe the communication technologies with necessary security features.

The major challenge was to bring the various technologies like Data Acquisition Systems, M2M Communications, Protocols, Physical-Cyber Security, etc. under a single umbrella which is in a succinct but palatable manner to the practicing Automation Engineers and Instrumentation Engineering students. Sincere attempt has been made to elaborate important topics appropriately while others are cited briefly. Each chapter is accompanied by a summary which will be useful for a quick revision. As the domain of expertise of most of the application engineers may be quite different, most of them may not be equipped enough to conceive the concepts of automation with SCADA because of the diversified requirements of ICT based systems for a successful implementation of Distributed Control Systems (DCS). Hence this book is written with an intention to bridge this gap to a considerable extent.

INTENDED AUDIENCE

This book is intended for a broad range of readers who will benefit from an understanding of Industrial Automation with SCADA architecture, secure Communication, SCADA Network Security, Attack Vectors in SCADA Systems, and associated technologies. This includes practicing Automation Engineers who are engaged with Industrial SCADA design and implementation, and Electrical, Electronics and Instrumentation engineering students who want to get involved in large scale automation. For the professionals who are already engaged in the design of Industrial SCADA with interoperable protocols, this book serves as a ready reference and is suitable for self-study. As an academic reference book, it is suitable for the senior under graduate and post graduate students. All the chapters of the book are very structured and modular to provide a considerable deal of flexibility for the design of courses in Industrial SCADA.

ORGANIZATION OF THE BOOK

This book has seven chapters covering the most important aspects of SCADA viz. Concepts, Communication and Security. However this book does not group chapters into parts. The first three chapters are dedicated to give brief but essential concepts of SCADA technologies with practical implementation solutions. The subsequent chapter gives an idea of various applications of SCADA. The chapters five and six are dedicated to modern SCADA communication technologies which revolutionized the SCADA. Much emphasis is given to make the automation engineers cognizant of the need, necessity and understanding of the modern communication technology and the communication protocols. The final chapter introduces the physical-cyber security aspects of the SCADA. Various cyber-attack types and vectors, and various methods to secure SCADA are discussed in this chapter.

ABOUT THE AUTHOR

The author holds a Masters in Electronics and Communication from Cochin University of Science and Technology (CUSAT) and another Masters in Solid State Electronics from University of Kerala with a graduateship in Electronics Engineering from Institution of Electronics and Telecommunication Engineers, New Delhi. After five years of initial Biomedical Research and Development assignment at Sree Chitra Tirunal Institute for Medical Sciences and Technology, Thiruvananthapuram with Dr.G.S.Bhuvaneswar who trained him to design and develop the hardware and software of SCADA systems, he has been working as an Electrical Engineer at various capacities in KSEB Ltd for the last 21 years during which he has been instrumental in making award winning model electrical distribution section, customizing and implementing the Power System SCADA/DMS project especially the design of the physical-cyber security aspects of automated power systems and Smart Grid. The implementation of SCADA/DMS project includes deployment and configuration of Firewalls, ensuring the End Node Security (ENS), High Availability (HA) etc. Beyond work he has been engaged into the vibrant and fascinating world of the creativity and chaos of the cosmic energy by attempting diverse forays and always toeing a fine line between physics and engineering since post-graduation.

CHAPTER WISE DESCRIPTION

CHAPTER 1

This chapter gives an introduction to SCADA with an emphasis on Data Acquisition Systems (DAS) and its components. The objectives and advantages as well as the evolution of SCADA are briefly discussed but with clarity. A brief discussion of the various components of DAS such as Sensors, Signal conditioners, Sample and Hold circuits, Analog to Digital Converters, etc. are also given in this chapter. Finally this chapter ends with an elaboration of the selection criteria of the DAS.

CHAPTER 2

This chapter begins with a comprehensive description of Remote Terminal Units, Programmable Logic Controller (PLC) and the different basic components of SCADA such as Intelligent Electronic Devices, Data Concentrator Units, Merging Units, Human Machine Interface, etc. The brief introduction of Data Concentrators and Merging Units are presented in such a manner that how digital substations can be designed. This chapter then gives an introduction of architecture of SCADA Master Stations and its hardware and software components. It concludes with a brief description of Geographical Positioning System (GPS), Situational Awareness and Alarm Processing.

CHAPTER 3

This chapter begins with describing the communication architecture of basic SCADA. Then moves on to describing the common communication philosophies adopted in DCS. As the reliability and availability of DCS functions are the most important, they are briefly introduced, but cater the necessary understanding to a power system professionals and engineering students who are engaged or intend to embark into the DCS, Smart Grid or Microgrid domain. It then explains the concepts of Fault Tolerant Systems, Fail Safe and redundant systems, High Availability, etc. Based on these concepts, this chapter then elaborates the design of a typical SCADA MCC architecture having redundant and HA connectivity.

CHAPTER 4

This chapter begins with a detailed description of the SCADA applications in power sector viz. the Energy Management System and Distribution Management System. Then it gives a brief description of SCADA applications in water pumping, water distribution, and water treatment. Attempts have also been made to briefly explain the SCADA applications in public transportation, automobile, and oil and gas industry.

CHAPTER 5

This chapter has focused on various communication aspects of the industrial SCADA with an emphasis on DCS and Smart Grid. It begins with discussing various types of transmission technology in very modest way so that it is very apt and most essential for a power engineer who is engaged in the design and implementation of SCADA and Smart Grid. The chapter then discusses the guided and unguided media used today for communication in such a manner that it is very useful for a practicing communication professional, which includes the various cabling issues as well. The various but most relevant communication technologies which find space not only in industrial SCADA but also in other smart automation technologies today are discussed comprehensively. Finally the chapter focused on the security issues of the wireless communication technology.

CHAPTER 6

This chapter begin with the evolution of communication protocols and then move on to explain the various communication protocols used today such as DNP3, Modbus, Profibus, IEC 60870, IEC 61850, and ICCP TASE 2 (IEC 60870-6). Other relevant ICS and PSS protocols such as IEEE C37.118.1 Synchrophasor Measurement Standard, IEC 61968 standard, IEC 61970 standard, IEC 62325 standard, IEC 61508, IEC 62351, IEC 62056 and IEC 62056-21 are also fleetingly presented. Finally the important points to select the right protocols are also described.

CHAPTER 7

This chapter gives a description about how the SCADA security is different from IT security and its importance. The requirement of open communication system and standardization are discussed with emphasis on security. Security concerns of the substation automation and control center architecture are discussed with solutions. Malware threats especially the attack of the lethal malware Stuxnet is a nightmare for SCADA implementing agencies because of the ZDVs of the Windows OS. These attack vectors and proposed solutions are discussed including threats and vulnerabilities of ICS and power system SCADA with various type of attacks and mitigating techniques are also discussed. Various proposals of the security engineers for making the automated power system smarter than cyber-attacks also have been elaborated.

SALIENT FEATURES

This book brings together timely and comprehensive information needed for an Automation Engineer to work in the challenging and changing area of Industrial Automation.

The book

- ⋏ Gives a deep understanding of the present Industrial SCADA technology.

- ⋏ Provides a comprehensive description of the Data Acquisition System to boost up the SCADA fundamentals and advanced communication technologies which revolutionized the Industrial Automation.

- ⋏ Imparts an essential knowledge of SCADA protocols used in industrial automation.

- ⋏ Comprehensive coverage of cyber security challenges and solutions of Industrial SCADA and Smart Grid which are the most essential topics for a practicing Automation Engineer who implements Distributed Control Systems.

- ⋏ Covers the state-of-the-art secure Communication technologies, key strategies, SCADA protocols, and deployment aspects in detail.

- ⋏ Enables practitioners to learn about upcoming trends, scientists to share new directions in research, and government and industry decision-makers to prepare for major strategic decisions regarding implementation of a secure Industrial SCADA technology.

- ⋏ Acquaints the current and leading edge research on SCADA security from a holistic standpoint.

ACKNOWLEDGEMENTS

Writing a technical book in seclusion would be a fretting bustle. The book cannot tell you whether you are making sense and, if you are, whether you are getting that sense across the readers. Hence the accessibility of technically aware colleagues and friends who are willing to find time both to discuss complex technical matters and to read and comment on sections of this book, are of great advantage to the author. Experimenting, documenting, reading and gathering information for this book has been a remarkable experience, as it opens new horizons and perspectives. I am very grateful to all my colleagues and friends who helped with many technical debates and especially for reading and commenting on particular sections of this book. At this moment of completion, I do recognize and appreciate my wife Dr.Shreelekshmi, my sons Harishankar and Harikrishnan, for their support and timely remarks on the presentation of this material. I do express my profound gratitude for the spiritual and ethical support of Swami Rithmbarananda of Sivagiri Mutt without which this attempt would not have been materialized. I also remember my parents with respect and my cousins especially V.R.Rana, Saji Natarajan, R.Ranjith, and K.S.Jayamohan, for all their sincere love and support which have also been very crucial in non-academic aspects. Beyond all these, special thanks to my classmate and friend Smt.T.G Parvathy who carried out the editing with many constructive suggestions and inspiring debates which helped me to finish the book in time. Finally I dedicate this effort with great respect to Prof.S.Suryadas, Sree Narayana College, Kollam, Kerala who introduced me to the challenging and exciting field of Electronics and Communication.

My sincere thanks to Notion team especially to Smt.Sneha Mathew and her colleague Smt. Agantuk Panwar, Publishing Manager for their support and co-operation on this project.

The author will be definitely privileged, if the readers get the intended sense which is the sole aim of this effort. Any suggestions for improvement of this book are always welcome.

- K S Manoj

LIST OF ACRONYMS

ADC	Analog to Digital Converter
ADU	Application Data Unit
AGC	Automatic Generation Control
AI	Analog Input
AMI	Advanced Metering Infrastructure
AMPS	Advanced Mobile Phone System
AMR	Automatic Meter Reading
ANSI	American National Standard Institute
AO	Analog Output
AP	Access Point
APDV	Application Protocol Data Unit
API	Application Program Interface
ARP	Address Resolution Protocol
ASCII	American Standard Code for Information Interchange
ASDU	Application Service Data Unit
BLT	Bilateral Tables
BNC	British Naval Connector
BPL	Broadband Over Powerline
CATV	Coaxial Cable TV
CCF	Common Cause Failure

CDMA	Code Division Multiple Access
CFE	Communication Front End
CIM	Common Information Model
CIS	Customer information System
CMF	Common Mode Failure
COSEM	Companion Specification for Energy Metering
CPU	Central Processing Unit
CSMA	Carrier Sense Multiple Access
D-AMP	Digital AMPS
DAS	Data Acquisition System
DC	Data Concentrator
DCS	Distributed Control System
DCU	Data Concentrating units
DI	Digital Input
DLMS	Device Language Message Specification
DMS	Distribution Management System
DMZ	De Militarized Zone
DNL	Differential Non-linearity
DNP3	Distributed Network Protocol 3.0
DO	Digital Output
DP	Decentralized Peripherals
DPM	Digital Panel Meter
DSL	Digital Subscriber Line
DSM	Demand Side Management
DSSS	Direct Sequence Spread Spectrum
DTS	Dispatcher Training Simulator

EDGE	Enhanced Data Rates for GSM Evolution
EIA	Electronic Industries Association
EMI	Electromagnetic Interference
EMS	Energy Management System
ENOB	Effective Number Of Bits
ENS	End Node Security
EPA	Enhanced Performance Architecture
EVC	Equipment Under Control
FDMA	Frequency Division Multiple Access
FEP	Front End Processor
FHSS	Frequency -hopping Spread Spectrum
FMS	Field bus Message Specifications
FRTU	Field Remote Terminal Unit
FTTH	Fiber To The Home
GOOSE	Generic Object Oriented Substation Event
GPRS	General Packet Radio Service
GPS	Global Positioning Systems
GSM	Global System for Mobile Communication
HA	High Availability
HDLC	High level Data Link Control
HFC	Hybrid Fibber Coaxial
HHU	Hand Held Unit
HMI	Human Machine Interface
HSDPA	High-speed Downlink Packet Access
ICCP	Inter Control centre Communication Protocol
ICS	Industrial Control System

ICT	Information and Communication Technology
IEC	International Electrotechnical Commission
IEDs	Intelligent Electronic Devices
IEEE	Institute of Electrical and Electronics Engineers
INL	Integral Non-linearity
IOT	Internet Of Things
IP	Internet Protocol
ISA	Industry Standard Architecture
ISDN	Integrated Service Digital Network
ISR	Information Storage and Retrieval
LAN	Local Area Network
LST	Low Side Transition
LTE	Long-Term Evolution
MAC	Media Access Control
MBP	Manchester Bus Powered
MIMO	Multiple Input and Output Manner
MITM	Man In The Middle attack
MMS	Multimedia Messaging Service
MU	Merging Unit
NC	Normally Closed
NEMA	National Equipment Manufacturers Association
NERC CIP	North American Electric Reliability Corporation - Critical Infrastructure Protection
NIST	National Institute of Standards and Technology
NMS	Network management Server
NMT	Nordic Mobile Telephone

NO	Normally Open
NSPF	No Single Point Of Failure
NVRAM	Non-volatile Memory
OFDMA	Orthogonal Frequency Division Multiple Access
OSI	Open System Interconnection
OTS	Operator Training Simulator
PA	Process Automation
PAC	Programmable Automation Controller
PCI	Peripheral Component Interconnect
PCMCIA	Personal Computer Memory Card International Association
PDC	Personal Digital Cellular
PDV	protocol Data Unit
PLC	Programmable Logic Controller
PLCC	Power Line Carrier Communication
PMU	Phasor Measurement Unit
PPP	Peer to Peer Protocol
PSS	Power System Scada
PSTN	Public Switched Telephone Network
RF	Radio Frequency
ROCOF	Rate of Change Of Frequency
RTUs	Remote Terminal Unit
RVDU	Remote Video Display Unit
SA	Substation Automation
SCADA	Supervisory Control And data Acquisition
SEP	Smart Energy Profile

SG   Smart Grid

SIL   Safety Integrity Level

SMSs   Short Message Service

SOE   Sequence of Events

SRLDC   Southern Regional Load Despatch Centre

TACS   Total Access Communication System

TCP/IP   Transmission Control Protocol/Internet Protocol

TDMA   Time Division Multiple Access

THD   Total Harmonic Distortion

TIA   Telecommunications Industry Association

UDP   User Datagram Protocol

UMTS   Universal Mobile Telecommunication System

UTP   Unshielded Twisted Pair

VDU   Visual Display Unit

VPN   Virtual Private Network

VPS   Video Projection System

WAN   Wide Area Network

WECA   Wireless Ethernet Compatibility Alliance

WG   Working Group

WiMAX   Worldwide Interoperability for Microwave Access

LIST OF TABLES

LIST OF FIGURES

Chapter ONE

INTRODUCTION TO SCADA

1.1 INTRODUCTION

SCADA (Supervisory Control and Data Acquisition) has been around us since there have been control systems and extensively used for monitoring and controlling geographically distributed processes in a variety of industrial processes. The first SCADA systems have been employed only for data acquisition by means of panels of meters, lights and strip chart recorders. Until recently many of the SCADA related products are proprietary, and the knowledge of the components has been acquired by the personnel while operating the system. Now the SCADA component manufactures started to follow standards which support interoperability. These help SCADA professionals to understand and design the SCADA systems in a systematic and structured manner. In the succeeding sessions an attempt has been made to elaborate on the essential components of the Data Acquisition Systems (DAS) with an emphasis on Distributed SCADA Systems (DCS) and Power System SCADA (PSS).

1.2 DATA ACQUISITION SYSTEMS (DAS)

A DAS is a system which gathers the input data in digital form accurately, quickly and economically as required. This consists of sensors with suitable signal conditioning, data conversion, data processing, multiplexing, data handling, associated transmission, and storage and display systems.

1.2.1 OBJECTIVES AND ADVANTAGES

The basic objectives of the Data Acquisition System are briefly described below.

1. Acquiring the required data reliably at required intervals.

2. The acquired data has to be appropriately processed and inform the status of the process/plant to the operator from time to time for controlling and decision making.

3. In general, DAS are designed to conceive a complete picture of the process/plant under monitoring intend to keep the plant/process operation safe and optimal.

4. With an effective HMI system or SCADA software, identify problem areas and minimize unit unavailability and maximize productivity at minimum cost.

5. DAS must be able to prepare sequence of events (SOE), summaries and store data for diagnosis, forecasting, etc.

6. Generally DAS are designed to compute unit performance indices using real-time data.

7. DAS must be flexible and capable of being expanded to accommodate future requirements.

8. DAS architecture design must be fail-safe, fault tolerant and reliable and down time must be less than 0.1%.

The major advantages of using DAS are increased reliability, lower operation and maintenance cost, faster restoration of the industrial process, reduction in human intervention and errors, accelerated decision making, with better accuracy, etc.

1.2.2 SINGLE CHANNEL DATA ACQUISITION SYSTEM

A Block Diagram of Single Channel Data Acquisition System is shown in Figure 1.1. The various components of the system are briefly explained below. It consists of Sensors, Signal Conditioner, Sample and Hold circuit,

Analog Digital Converter and a storage device/printer/PC. Followed by an Analog to Digital Converter (ADC), performing repetitive conversions at a free running, internally determined rate. The outputs are in digital code words including over range indication, polarity information and a status output to indicate when the output digits are valid.

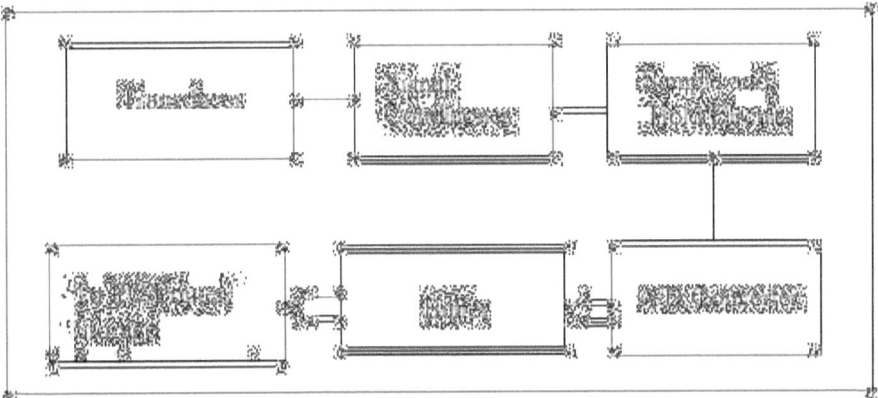

Figure 1.1   Single Channel Data Acquisition System

The digital outputs are further fed to a storage or printout device, or to a digital computer device for analysis. The popular Digital Panel Meter (DPM) is a well-known example of this. However, there are two major drawbacks in using it as a DAS. It is slow and the BCD has to be changed into binary coding, if the output is to be processed by digital equipment. While it is free running, the data from the A/D converter is transferred to the interface register at a rate determined by the DPM itself, rather than commands beginning from the external interface.

1.2.3   MULTI-CHANNEL DATA ACQUISITION SYSTEM

By suitably incorporating a Multiplexer, a multi-channel DAS can be designed. By suitably selecting the address of the MUX, the input channels can be polled depending upon the priority. A block diagram of a multi-channel DAS is shown in Figure 1.2. Multi-channel DAS is elaborated in subsequent chapters.

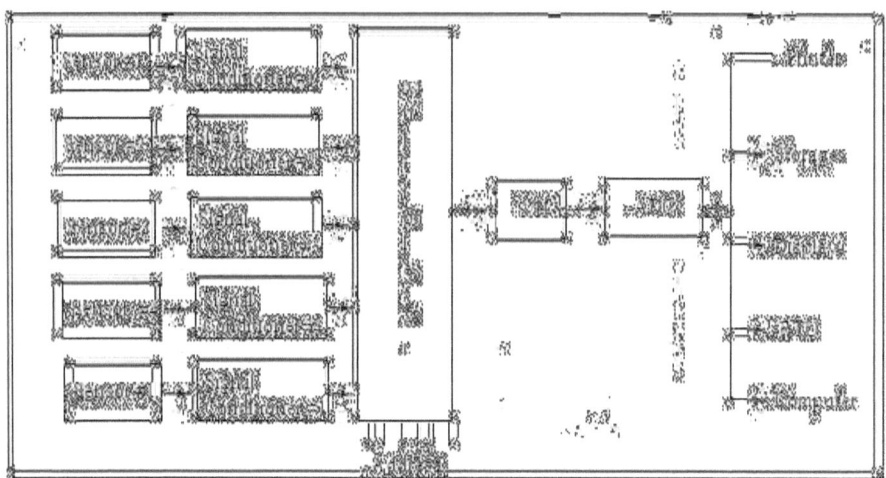

Figure 1.2   Multi-Channel DAS

When simultaneous measurements of the physical quantities are to be taken, the S/H circuit generally placed before MUX as shown in Figure 1.3. All the required parameters which are to be captured simultaneously are sampled and held at the same instance and then digitized one after the other with software polling.

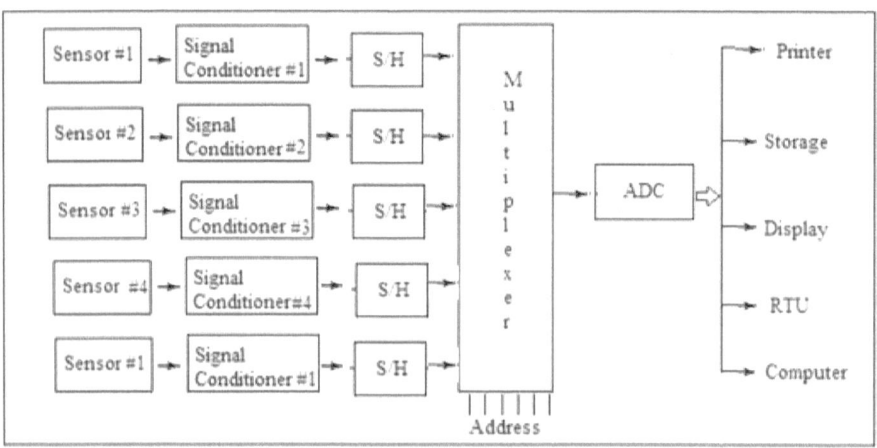

Figure 1.3   Multi-Channel DAS for simultaneous measurement

1.2.4 SENSORS

Sensors or Transducers are devices that detect and measure a physical quantity such as pressure, force, temperature, acceleration, etc. and provide a corresponding output in the form of an electrical signal like voltage, current, resistance or frequency. Transducer characteristics define many of the signal conditioning requirements of the measurement system.

1.2.5   SIGNAL CONDITIONING

In general the quality of the signals obtained from the sensors has to be enhanced appropriately to bring it to an acceptable level to the Analog to Digital Converters (ADC). This includes signal scaling, amplification, attenuation, linearization, filtering, anti-aliasing, excitation etc. Further signal conditioners have an additional responsibility to protect from unintentional or accidental high voltage inputs or surges, etc. Direct digital conversion carried out near the signal source is very advantageous in cases where data needs to be transmitted through a noisy environment. Even with a high level signal of 10 V, an 8 bit converter having a 1/256 resolution can produce 1 bit ambiguity when affected by noise of the order of 40 mV. Presently transducers being developed which is combined with ADC capable of converting to encrypt digital data.

Excitation: Not all sensors are active. In such cases, where the sensing devices are passive, it requires a voltage or current excitation and is supplied by the signal conditioners.

Amplification and Attenuation: If the signal acquired is large, then a simple attenuator, is used to scale down the input gains, in order to make it acceptable to the input signal range of the ADC. However most of the sensors generate the signals of low amplitude of voltage, current, or resistance, etc. In this case an amplifier circuit of suitable gain is employed to bring them to the acceptable level. If the sensor output is in the form of change in resistance, then a bridge circuit is most ideal to detect the change in resistance. A bridge amplifier is most suitable for amplifying the bridge outputs and improving the sensitivity of detection.

Isolation: When involving with high voltage it has to be ensured that the signals are physically isolated between the sensor output and the rest of the system. It is achieved by breaking the conductor paths by magnetic coupling, optical coupling, or by capacitive coupling. Optical coupling uses an LED at the transmitting side and a photo diode at the receiving side. Capacitive coupling uses a capacitor to isolate input and output signals. Magnetic coupling uses a transformer to isolate the input and output.

These techniques are advantageous in handling signals from high voltage sources and transmission towers. In biomedical applications such isolation is unavoidable.

Linearization: Sensors which gives non-linearization response can be corrected by proper signal conditioning techniques. Linearization of the data, can be performed by analog techniques using either linear-approximation, or smooth series approximation using a low cost IC amplifier. Alternately linear approximation can be performed digitally after data acquisition and conversion by the use of ROMs by storing a suitable linearization table or programme initially.

Filtering and Anti-Aliasing: As most of the sensor output level is very low, they are very prone to Electro Magnetic Interference. Appropriate filters are used to eliminate the noise from the signals. Table 1.1 summarizes the basic characteristics and signal conditioning requirements of typical transducers.

Table 1.1   Sensors with signal conditioning requirements

1	Thermocouple	Amplification, Linearization and Reference temperature for cold junction compensation.
2	Strain Gauge	Required excitation voltage or current, Bridge formation, amplification, and linearization.
3	LVDT	Excitation and linearization.
4	RTD	Current excitation and linearization
5	Thermistor	Current or voltage exciter and linearization

1.2.6   SAMPLE AND HOLD CIRCUIT

To achieve a reliable and accurate analog to digital conversion, ADCs require a fixed time during which the input signal remains constant called aperture time. This is a requirement of the conversion algorithm used by the A/D converter. If the input changes during this time, the ADC output will be inaccurate. This situation can be managed by suitably incorporating a sample-and-hold device. It samples the output signal from the multiplexer or gain amplifier very quickly and holds it constant for the ADCs aperture time. Usually sample and hold circuit is placed between multiplexer and ADC.

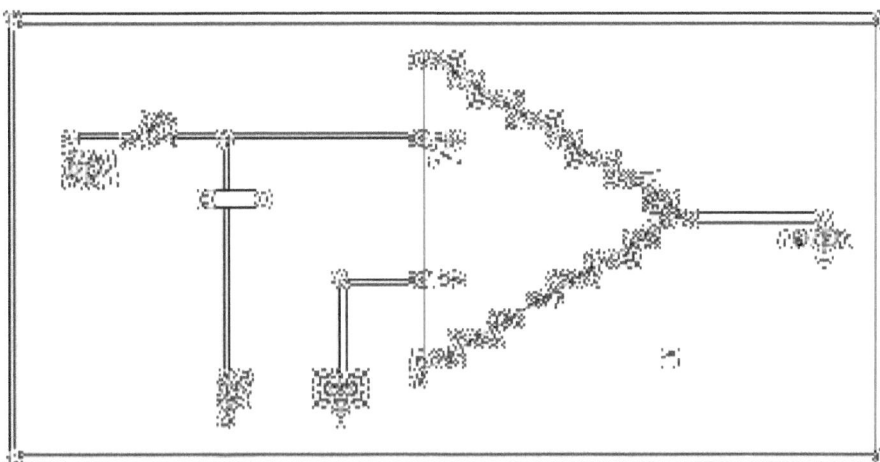

Figure 1.4   Sample and Hold circuit

Sample-and-hold circuit can be approximated by a capacitor and a high gain opamp. A typical sample-and-hold using an opamp is shown in Figure 1.4. When the switch is closed, the capacitor charges to the input voltage. When the switch is opened, the capacitor holds the voltage level until the next sampling time. The opamp provides large input impedance so that the capacitor is not discharged appreciably and at the same time offers the gain to drive external circuit.

1.2.7   A TO D CONVERTERS (ADC)

An analog-to-digital converter, or ADC, is a device or peripheral that converts analog signals into digital signals. In the real world, signals mostly exist in analog form. An ADC with a S/H circuit can be used to sample such signals and the signals can be converted to the digital values. There are four types of A/D converters generally used, and they are

1.   Integrating or dual slope ADC,

2.   Successive approximation ADC,

3.   Parallel Comparator (Flash) type ADC, and

4.   Counting type ADC.

1.2.7.1   Integrating or Dual Slope ADC

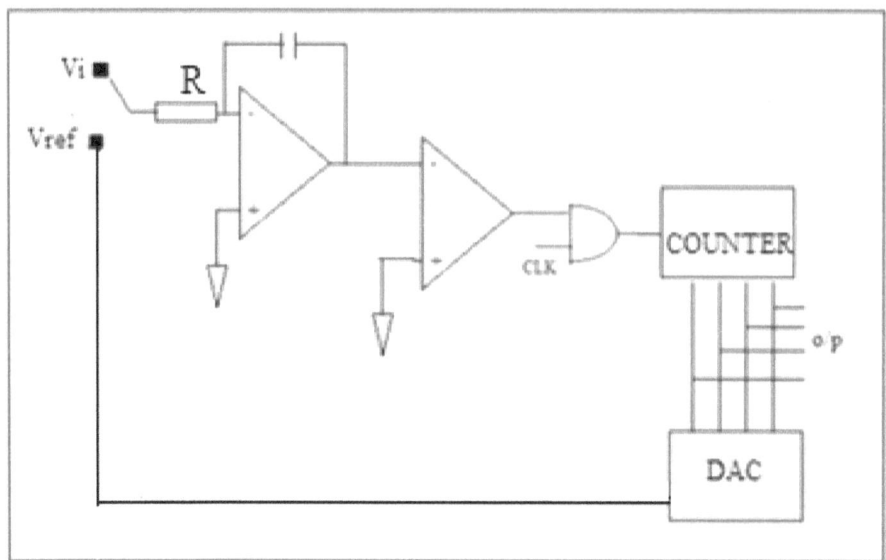

Figure 1.5   Integrating or Dual Slope ADC

These are used for very low frequency and may have very high accuracy and precision. They are found in thermocouple and RTD modules. Other advantages include very low cost, relatively less noise and mains pickup tend to be reduced by the integrating and dual slope nature of the A/D converter. The A/D procedure essentially requires a capacitor to be charged with the input signal for a fixed time, and then uses a counter to calculate how long it takes for the capacitor to discharge. This length of time is proportional to the input voltage. It is more accurate ADC type among all. It has greater noise immunity compared to other ADC types. However, it is the slowest ADC among all.

1.2.7.2 Successive Approximation ADC

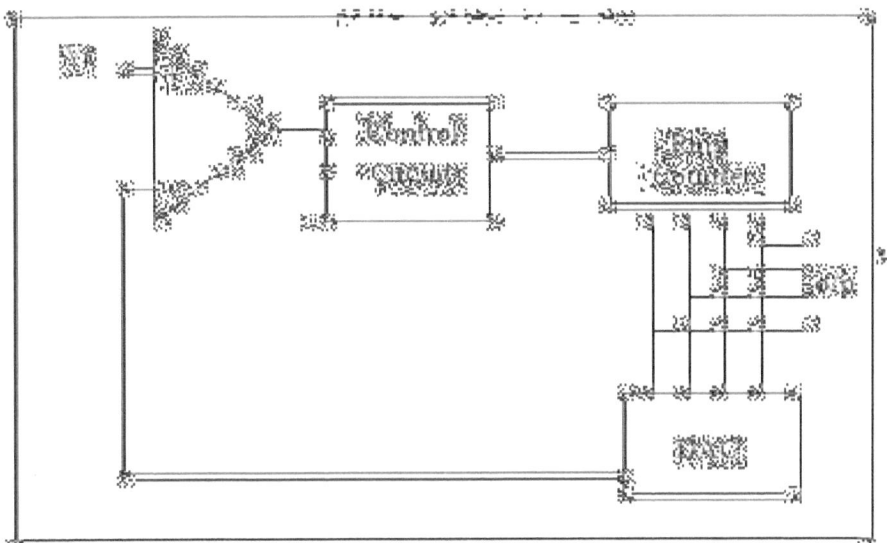

Figure 1.6 Successive Approximation ADC

Successive approximation A/Ds allow much higher sampling rates (up to a few hundred kHz with 12 bits is possible) while still being reasonable in cost. The conversion algorithm is similar to that of a binary search, where the ADC starts by comparing the input with a voltage (generated by an internal DAC converter), corresponding to half of the full-scale range. If the input is in the lower half, the first digit is zero and the ADC repeats this comparison using the lower half of the input range. If the voltage had been in the upper half, the first digit would have been 1. This dividing of the remaining fraction of the input range in half and comparing to the input voltage continues until the specified number of bits of accuracy has been obtained. It is obviously important that the input signal does not change when the conversion process is underway.

1.2.7.3   Parallel Comparator (Flash) ADC

It is the fastest ADC among all the ADC types. In flash ADC of n-bit in size, (2n - 1) comparators and 2n registers are needed. Each comparator compares Vin to a different reference voltage, starting with Vref = 1/2 (LSB). Op-amp is used as comparator here. Figure 1.7 depicts block diagram of parallel comparator ADC. It is very fast as mentioned. But hardware requirements are very high. As an example, 255 comparators are required for an 8-bit ADC. Low resolution and large power consumption are the major disadvantages.

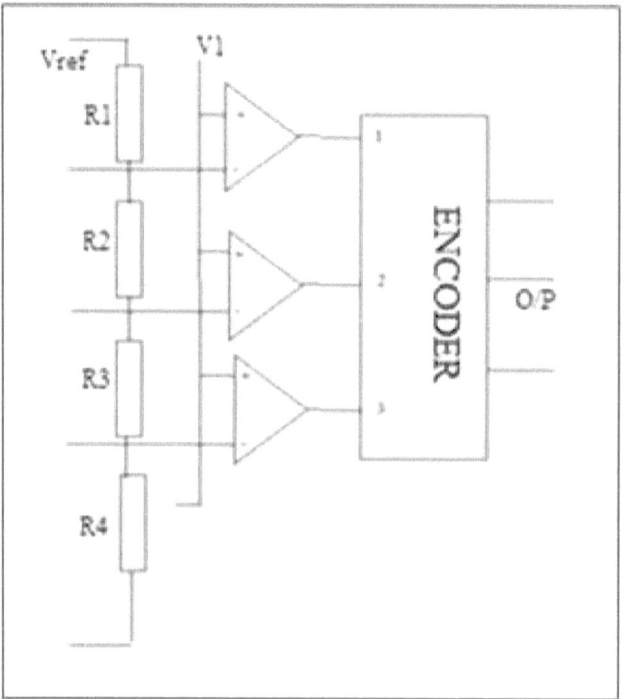

Figure 1.7   Parallel Comparator (Flash) type ADC

1.2.7.4   Counting Type ADC

Counting type ADC uses counter and DAC. It compares DAC output with analog voltage and does the same till both are equal in magnitude. At this moment counter will stop. The conversion time depends on analog input voltage. Figure1.8 depicts block diagram of counter type ADC.

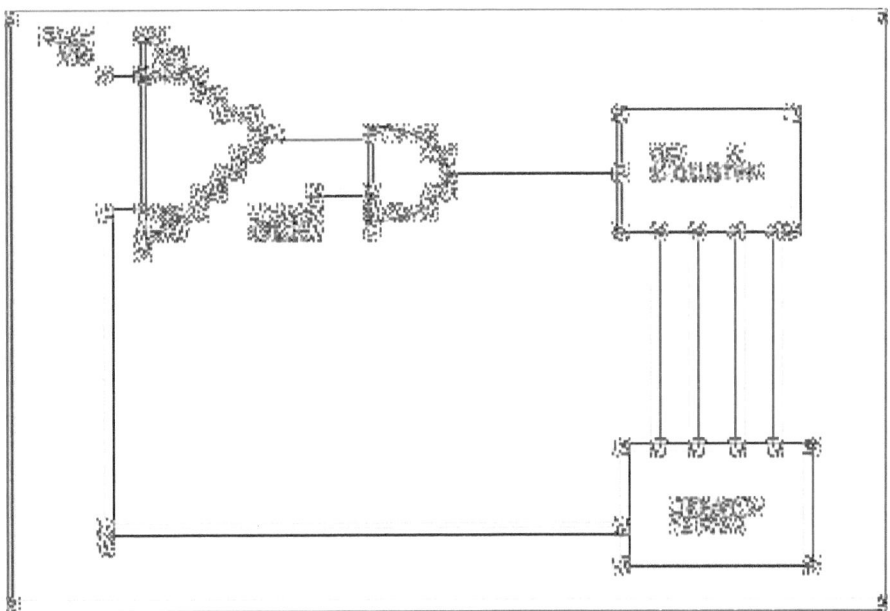

Figure 1.8 Counter Type ADC

1.2.7.5 ADC Specifications

Certain important specifications of A to D converters are briefly explained below.

Accuracy: It includes quantisation error, system noise, linearity, etc. Absolute accuracy refers to the maximum analog error.

Common Mode Rejection Ratio (CMMR): This is the ratio of the resulting output signal to a changing input common-mode signal. It is the degree of the rejection of a common mode signal across the differential output stage.

Conversion Time: The conversion time is the time required for an ADC to complete the single conversion. This time, does not include the acquisition time and MUX time. The conversion time for a given ADC is preferably less than the throughput time.

Crosstalk: In the multi-channel DAS, coupling between the adjacent channels, and sharing the transmission path, results in crosstalk. This interference appears as a noise in the digital output and quite unwanted.

Input Range: The specified range of the peak to peak, input signal of an A/D converter.

Latency: Latency is the time required for an ideal step input to converge, within an error margin to a final digital output value. The error-band is expressed as a predefined percentage of the total output voltage step. The latency of a conversion is that period between the time where the signal acquisition begins to the time to the next conversion starts.

Linearity Errors: With most ADCs gain, offset and zero errors are not critical as they may be calibrated out. Linearity errors, differential non-linearity (DNL) and integral non-linearity (INL) are more important because they cannot be removed.

Differential Non–Linearity (DNL):   It is the difference between the actual code width from the ideal width of 1 LSB. If DNL errors are large, the output code widths may represent excessively large and small input voltage ranges. If the magnitude of a DNL is greater than 1 LSB, then at least one code width will vanish, yielding a missing code.

Integral Non-Linearity (INL):   INL describes the non-linearity of ADC. It is considered as an important parameter because it is a measure of an ADC non-linearity error. However, as in any Analog or Mixed-Signal Design project, some specifications are important, some are not. It all depends on the project requirements regarding accuracy and precision. Understanding INL enables the circuit designer to avoid surprises in his or her project. INL is defined as the maximum deviation of the ADC transfer function from the best-fit line.

Resolution: Resolution can be used to describe the general performance of an ADC. It is the smallest change that can be distinguished by an ADC converter. For example, for a 12-bit ADC converter resolution would be 1/4096 = 0.0244%. It's an important ADC specification because it determines the smallest analog input signal an ADC can resolve.

Monotonicity: This requires a continuously increasing output for a continuously increasing input over the full range of the converter. This term implies that an increase (decrease) in the analog voltage input will always produce no change or an increase in the digital code.

Quantizing Uncertainty: Because the ADC can only resolve an input voltage to a finite resolution of 1 LSB, the actual real-world voltage may be up to ½ LSB below the voltage corresponding to the output code or up to ½ LSB above it. An ADC's quantizing uncertainty is therefore always ±½ LSB.

Relative Accuracy: This refers to the input to output error as a fraction of full scale with gain and offset error adjusted to zero.

1.2.8   STORAGE AND DISPLAY

Once the analog data is converted into digital format, it can be suitably stored, displayed, processed, or send to remote master stations through an appropriate communication medium.

1.2.9   DATA FORWARDING AND COMMUNICATION

In certain occasions, the information acquired has to be send to remote locations, where other SCADA components or master stations are situated. This is general requirement in the case of Distributed SCADA such as Power System SCADA (Energy Management System and Distribution Management System). This arises the requirement for an appropriate communication technology which can provide a secure communication. Today a variety of communication technologies suitable are available which will be discussed in the subsequent chapters.

1.3   EVOLUTION OF SCADA

SCADA systems can be considered that, it has been evolved through four generations as described below.

1.3.1   MONOLITHIC SCADA – THE FIRST GENERATION SCADA

These SCADA systems were in use before the revolution of computer networking. They were stand-alone systems having no connectivity to other systems and developed as local SCADA. These mostly used guided communication with proprietary protocols. Further these systems did not envisage the fail safe and fault tolerant design aspects seriously. Hence these SCADA systems faced reliability issues considerably. The hardware used was mainly minicomputers capable of large computing capabilities.

1.3.2   DISTRIBUTED SCADA – THE SECOND GENERATION SCADA

This generation came into existence when the computer networking technology has been developed and incorporated into the SCADA systems. Here the data gathered and the data processing across the multiple stations are connected through computer networking. The latency has been reduced considerably to an extent that system as a whole functioned in near real-time. These SCADA system were economical when compared with first generation, as each station has been assigned with particular tasks. However the protocols used was still proprietary and was not interoperable. But the proprietary protocols had an advantage that beyond the developers, the extent of security or the security flaws are unknown. In other way, these SCADA systems were secured through *security through obscurity.*

1.3.3   SCADA WITH STANDARD PROTOCOLS – THE THIRD GENERATION SCADA

The advancement of communication technology and with the introduction of interoperable SCADA protocols, SCADA broke the geographical barriers and spread across more than one LAN network called Process Control Network. The master station may have several servers running parallel to handle various tasks, such as historian, SCADA, NMS, Development, etc. This makes the system very economical and real-time. However the physical-security is a major concern and must be addressed while designing and implementing.

1.3.4   INTERNET OF THINGS – THE FOURTH GENERATION SCADA

By appropriately integrating the advancement of cloud computing, SCADA has taken a new shape and adopted the name Internet of Things. The advantages of amalgamating the various technologies are

1. Capable of being flexible and affordable with the ability to go private.
2. Capable of ingest massive amount of machine data
3. Capable of connecting different machines and systems such as SCADA, DCS, and historians.

4. Capable of connecting various machines having net connectivity and process data across all these sources together.

5. Capable of real-time complex and real-time processing of data from multiple sources.

6. Capable of Big data processing and apply supervised and unsupervised machine learning algorithms to predict outcomes.

Another advantage of incorporating the cloud computing technology is that it significantly reduces the infrastructure costs and increases the ease of maintenance. Further the SCADA operations become near real-time and the use of open protocols with TLS security improves the security boundary considerably.

In fact, the fourth generation SCADA or IoT transforms the human centric internet into objects or things centric. It is expected that about 70 billion things may be hooked into internet in the near future while people hooked into the internet may be only 7 billion. However at present, IoT is not a tangible reality but it is a prospective vision of a number of technologies. But once materialized, it can drastically change the way of functioning of our society. Realizing the potential and the opportunity of business, IoT has become a buzzword in many countries. Further it is anticipated that the whole world will be soon under the influence of IoT wave.

1.4   COMMUNICATION IN SCADA

The innovations of ICT in SCADA in fact, helped to overcome the geographical barriers and made the remote monitoring and control an easy task. The main communication requirements are between field devices and the RTU/PLC/DCU, and the communication between the RTU and the Master Control Center. Many proprietary and open communication protocols suitable for industrial SCADA have been developed. Today many open protocols developed are under modification to incorporate security features. A detailed description of the SCADA communication and communication protocols are given in the subsequent chapters.

1.5   SELECTION CRITERIA OF DAS

Today many companies are manufacturing the SCADA components using different technologies. Choosing the right SCADA system components from this wide array of SCADA products is a daunting task. As already explained SCADA component starts with data acquisition components, which translate the physical world analog and discrete signals to digital data suitable for processing by the digital computers. Hence the fundamental and prime component of the DAS is the ADC. Many DAS products today available in the market are with communication and control capabilities and having a range of I/O modules. Hence with an in-depth knowledge of the DAS specifications, one can fully optimize their requirements and select the most suitable and economical solution so that one can avoid paying for features which one don't require.

Input Range: Selection of input range is very important as it depends the nature of the physical quantity to be measured and the type of the transducer used. Usually DAS provide a range which matches the maximum output range of the transducer. This results in providing the highest possible resolution. Many DAS provide multiple input ranges by using software-programmable-gain amplifiers.

Number of input channels: The number of input channels on data acquisition boards typically ranges from 4 to 64. Input channels can be single-ended (SE) or differential (DI). Differential inputs offer noise immunity, and can improve accuracy when long cables, low-level input voltages (less than 1 V full-scale), or high-resolution converters are used.

Accuracy: The ADC of the DAS converts the analog value of the physical quantity to digital data. Accuracy has been defined by how closely the binary code matches the true value of the monitoring analog signal. The major component of accuracy, and its limiting factor, is resolution. Stated in bits, resolution determines the number of counts or binary numbers used to represent the analog signal. When evaluating a data acquisition board, it is critical to understand the relationship between resolution and accuracy. Resolution is simply one factor affecting accuracy.

Total harmonic distortion (THD):   It is the ratio of the sum of the harmonics of the fundamental frequency of the input signal to the fundamental frequency itself. THD is a good indicator of the quality of the circuit design. A high THD measurement indicates a flawed analog-input design.

Speed: The throughput of a board, generally specified in megasamples per second or kilosamples per second is a crucial measurement for high-speed applications. If multiple ADCs are used on a single board, the specified throughput represents the sum total of the individual converter throughputs. ADC throughput is mainly determined by three elements:

1. conversion time which is the time needed to do actual conversion,

2. acquisition time which is the time needed by associated acquisition circuitry-the multiplexer, signal conditioning, and sample and hold-to acquire a signal accurately, and

3. transfer time which is the time needed to transfer data from the board to system memory.

Some high-speed DAS boards increase throughput by overlapping the acquisition time on one sample with the analog-to-digital conversion time of the previous sample, in effect, handling two signals at one time. Most analog output circuits have a separate data buffer for each channel. Settling time varies proportionately with the size of the output change, and is specified in microseconds.

Clocks, Triggers, etc. With multiple pacer clock circuits, multiple conversions with precision timings suitable for time stamped data acquisition are possible. Pacer clock triggering is achieved by a software instruction or with a hardware digital pulse or with an analog voltage. Most of the RTUs and PLCs contain general purpose counter/ timer circuits for time stamped data acquisition. These consist of several counters and a frequency source. While confirming the pacer clock circuits in DAS, one should not be confused with the counter/ timers available on the board, which are dedicated to the clocking and triggering of the analog-to-digital and digital-to-analog subsystems.

In simple local SCADA systems, PCI data acquisition boards can feed acquired data directly to the PC's memory, eliminating the need for on-board memory and the resulting gaps in data. Furthermore, PCI boards are auto-configured upon installation to match the system's resources. Manual configuration of jumpers or DIP switches is a thing of the past. When frequency-domain performance is characterized, one can be sure that the acquired data will be as accurate as required it to be. The Effective Number of Bits (ENOB) specification is the way to clearly convey a data acquisition board's ac performance. Combining all critical, real-world performance considerations such as accuracy, settling time, and dynamic performance in one easy-to-comprehend specification, ENOB specifies the overall accuracy of the analog-to-digital transfer function.

Standards and Certifications: The FCC and CE certification standards are two good indicators of quality and reliability. These certifications guarantee the buyers that DAS components are meeting certain standards and will perform robustly in a real-world situation.

Planning For Future Changes: A data acquisition system should meet the present needs and provide flexibility for the future, while justifying the utilities investment in hardware and software. Common upgrades to a DAS is changing, especially the hardware and hence with new add-on boards. If the application software is not designed with an open systems approach, adding new boards can cause expensive, time-consuming reprogramming. Making sure that the Application Programming Interface (API) is hardware-independent will allow changing boards with significantly no or little reprogramming. Also, make sure the data acquisition software support multiple operating systems. Ensure that DAS software is supported for the upgraded version of Operating System. This will reveal the fact of difficulty of migration from one of the early versions of OS to the upgrades.

Another hardware issue involves the use of interrupts. Traditionally, computer peripherals request the host CPU's attention via the hardware interrupts on the CPU. However, the number of peripheral components on a typical system (modem, scanner, CD-ROM drive, etc.) has increased to the point where it frequently exceeds the fifteen interrupts available. To address this problem, Data Translation and some other board manufacturers have stopped using hardware interrupts on new boards, achieving the same function with a software feature.

Hence while planning a data acquisition system or SCADA for an utility, keep the application in mind, comprehend the requirements clearly and choose wisely the data acquisition system, the software and the communication technology which is relevant and most appropriate. Also look for simple solutions rather complex one but with accuracy and robustness. Also ensure that software is designed for future upgradability.

SUMMARY

This chapter gives an introduction to SCADA with an emphasis on Data Acquisition Systems (DAS) and its components. The objectives and advantages as well as the evolution of SCADA are briefly discussed but with clarity. A brief discussion of the various components of DAS such as Sensors, Signal conditioners, Sample and Hold circuits, Analog to Digital Converters, etc. are also given in this chapter. Finally this chapter ends with an elaboration of the selection criteria of the DAS.

Chapter TWO

SCADA SYSTEM COMPONENTS

2.1 INTRODUCTION

All SCADA or Distributed Control System (DCS) starts with the field equipment depending on the process/plant to be monitored and controlled. The appropriate sensors pick up the process parameters and convert into proportional electrical voltage or current signal. These electrical signals are then conditioned and converted as per the requirement into digital form by means of an ADC. This electrical signal in digital format is then communicated to the SCADA server through different devices and protocols depending upon the SCADA architecture and the communication technologies employed. This chapter elaborates various components such as Remote Terminal Unit (RTU), Programmable Logic Controller (PLC), Merging Units (MU) and Data Concentrators (DC), Master Control Station and SCADA Servers, HMI, etc. of the modern industrial SCADA environments.

2.2 REMOTE TERMINAL UNIT (RTU)

Remote Terminal Unit (RTU) is a microprocessor based device connected to appropriate sensors, transmitters and process equipment for the purpose of remote telemetry and control of plant or process. RTUs find applications in oil and gas remote instrumentation monitoring, networks of remote pump stations, environmental monitoring systems, air traffic equipment, power utilities, etc. RTUs with the aid of appropriate sensors, monitors production processes at remote site and transmits all data to a central station where it is gathered and monitored. An RTU can be interfaced using serial ports such as RS232, RS485, and RS411 or Ethernet to communicate with the central stations. They also support various protocol standards such as Modbus, IEC 60870, DNP3 making it possible to interface with 3rd party software.

2.2.1   EVOLUTION OF RTUs

The beginning of the last century witnessed many attempts to automate and remotely control many plants and process which resulted in the development of a variety of automation systems. However the systems developed at that time could only monitor and was not capable of controlling effectively. Subsequent innovations by the automation engineers made it possible to convey the status change of a process to a remote monitoring station and print the status change with reported time and date.

With the development of solid state technology, electromechanical systems were slowly replaced with the solid-state components such as electronic sensors, and analog-to-digital converters. The advent of microprocessors, equipped RTU manufacturers to upgrade their technology by incorporating microprocessor and microcomputer logic systems into RTUs. This enhanced the flexibility of Industrial Control Systems (ICS) by bringing in various new capabilities in operation and performance. The development in communications and faster high computing microprocessor chips brought down the costs and improved the performance of ICS. The microcomputer based new systems have the following major benefits.

1.   Suitable for modular system design,
2.   Mostly available with a pre-programmed HMI system,
3.   User friendly menu-driven HMI software,
4.   High bandwidth communication capabilities,
5.   A variety of in-built diagnostic strategies,
6.   Easy maintenance and replacement of circuit boards,
7.   Redundancy at all level to ensure the high availability and improve the reliability, and
8.   Supports most of the standard communication protocols.

The RTUs with a wide range of functionalities, such as secure remote access and communication, capable of ensuring the end node security, optional local data processing and decision making capabilities to shed processing burden of Master Control facility are presently available.

2.2.2   RTU ARCHITECTURE

The RTU architecture comprises of a CPU, volatile memory and non-volatile memory for processing and storing programs and data. It communicates with other devices via either serial ports or an on-board modem with I/O interfaces. It has a power supply module with a backup battery, surge protection against spikes, real-time clock and a watchdog timer to ensure that it restarts when operating in the sleep mode. Figure 2.1 shows the block diagram of an RTU configuration. A typical RTU hardware module includes a control processor and associated memory, analog inputs, analog outputs, counter inputs, digital inputs, digital outputs, communication interfaces and power supply. Centralized RTU design where all I/O modules are housed in RTU panels and communicating with master station through communication port.

Distributed RTU design in which distributed I/O modules or processor with I/O modules are housed in respective RTU panel. All these distributed I/O modules shall be connected to a central processor for further communication with master station. The customer shall asses the requirement of RTU panels for such design and supply panels accordingly.

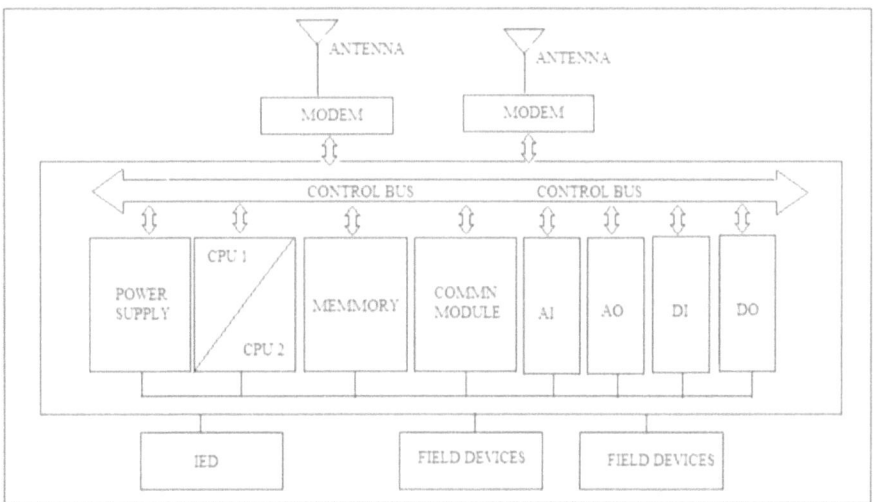

Figure 2.1   RTU Architecture

2.2.2.1   Central Processing Unit (CPU)

Today RTU designs utilize a 16 bit or 32 bits microprocessor with a total memory capacity of 256kbytes expandable to 4 Mbytes. It also has two or three communication ports or multiple Ethernet links. This system is controlled by a firmware and a real-time clock with full details of day, month and year is used for accurate time stamping of events. A watchdog timer provides a check that the RTU program is executing regularly. The RTU program regularly resets the watchdog timer and if this is not done within a certain time-out period the watchdog timer flags an error. Depending on the requirements of safety, reliability, high availability, and functionality, dual CPU with dual power supplies are often preferred. The systems provide for automatic monitoring of primary and hot standby CPUs. Failures are detected automatically and trigger a switch over from the primary CPU to hot standby CPU.

2.2.2.2   Analog Input Modules (AI)

An analog input signal is generally a voltage or current that varies over a defined value range, in direct proportion to a physical process measurement. 4-10 milliamp signals are most commonly used to represent physical measurements like pressure, flow and temperature. Analog inputs of different types including 0-1 mA, 0–10 V, ±1.5 V, ±5.0 V etc. are also common and acceptable to RTU. Five main components of the analog input (AI) module are described below.

1. *Multiplexer:* This samples several analog inputs in turn and switches each to the output in sequence. The output goes to the analog digital converter.

2. *Signal Conditioner:* This amplifies and transforms the low-level voltages to match the input range of the board's A/D converter

3. *Sample and hold circuit:* An analog device that samples the voltage of a continuously varying analog signal and holds its value at a constant level for a specified minimum period of time

4. *A/D converter:* A system that converts an analog signal into a digital format which is a digital code corresponding to the input voltages.

5. Bus interface and board timing system.

Typical analog input modules have the following features:

1. 8, 16, or 32 analog inputs

2. Resolution of 8 to 12 bits

3. Range of 4-10 mA

4. Input resistance typically 140kohms to 1 MΩ condition and can sometimes reset the CPU.

5. Conversion rates typically 10 microseconds to 30 milliseconds.

RTU can also receive analog data via a communication system from a master or Intelligent Electronic Device (IED) sending data values to it.

2.2.2.3   Analog Output Module (AO)

Though it is not commonly used, analog output (AO) modules are included in control devices to deal with varying quantities, such as graphic recording instruments or strip charts. The function of an Analog Output module is to convert a digital value supplied by the CPU to an analog value, by means of a digital to analog converter (DAC). This analog representation can be used for variable control of actuators. The basic features of the Analog output modules are as follows.

1. 8, 16 or 32 analog outputs

2. Resolution of 8 or 12 bits

3. Conversion rate from 10micro seconds to 30 milliseconds

4. Outputs ranging from 4-10 mA or 0 to 10 volts

2.2.2.4   Digital or status inputs (DI)

These are used to indicate status and alarm signals. Most RTUs incorporate an input section or input status cards to acquire two state real world information. This is usually accomplished by using an isolated voltage or current source to sense the position of a remote contact (open or closed) at the RTU site. This contact position may represent many different devices, including electrical breakers, liquid valve positions, alarm conditions, and mechanical positions of devices.

2.2.2.5 Digital Output Modules (DO)

RTUs may drive high current capacity relays to switch power on and off to devices in the field. The DO modules are used to drive an output voltage at each of the appropriate output channels with three approaches possible.

1. Triac Switching,
2. Read Relay Switching, and
3. TTL voltage outputs

2.2.2.6 Power Supply Module

RTUs need a continuous power supply to function, but there are situations where RTUs are located at quite a distance from an electric power supply. In these cases, RTUs are equipped with alternate power source and battery backup facilities in case of power losses. Solar panels are commonly used to power low-powered RTUs, due to the general availability of sunlight. Thermo electric generators can also be used to supply power to the RTUs where gas is easily available like in pipelines. Normally RTU is expected to operate from 110/140 V AC ± 10% 50 Hz or 11/14/48 V DC± 10% typically. Batteries that should be provided are lead acid or nickel cadmium. Typical backup requirements are for 10-hour standby operation and a recharging time of 11 hours for a fully discharged battery at 15°C. The power supply, battery and associated charger are normally contained in the RTU housing. The monitoring parameters of the battery system of RTU which should be transmitted back to the central site/master station are analog battery reading and alarm for battery voltage outside normal range

2.2.2.7 Communication interfaces

Modern RTU are designed to be flexible enough to handle multiple communication media such as

1. RS 232/RS442/RS 485 etc.
2. Ethernet
3. Dial up telephone lines/dedicated landlines

4. Microwave, and Satellite,

5. X.15 packet protocols, and

6. Radio via trunked/VHF/UHF.

An RTU may be interfaced to Multiple Control Stations and Intelligent Electronic Device (IEDs) with different communication media such as RS232, RS485, etc. An RTU may support standard protocols (Modbus, IEC 60870-5-101/103/104, DNP3, IEC 60870-6-ICCP, IEC 61850 etc.) to interface any third party software.

Data transfer may be initiated from either end using various techniques to insure synchronization with minimal data traffic.

There are two methods in general viz. polling and report by exception. In polling method, the master may poll its subordinate unit (Master polls RTU or RTU polls IED) for changes of data on a periodic basis.
The report by exception method is used where a subordinate unit initiates an update of data upon a predetermined change in analog or digital data. Periodic complete data transmission must be used periodically, with either method, to insure full synchronization and eliminate stale data. Most communication protocols support both methods, programmable by the installer. Analog value changes will usually be reported only on changes outside a set limit from the last transmitted value. Digital Status values observe a similar technique and only transmit groups (bytes) when one included point (bit) changes.

Multiple RTUs or multiple IEDs may share a communications line, in a multi-drop scheme, as units are addressed uniquely and only respond to their own polls and commands. IED communications tranfer data, between the RTU and an IED. This can eliminate the need for many hardware status inputs, analog inputs, and relay outputs in the RTU. Communications are accomplished by copper or fibre optics lines. Multiple units may share communication lines.

Communications to an MCC are generally envisaged and suitably incorporated in larger systems. The communication media used to transfer data may be copper, optical fibre or wireless media communication system. Multiple units usually share communication channels.

2.2.3   RTU ENVIRONMENTAL ENCLOSURES

Typically, the printed circuit boards of DI, AI, AO, DO, etc. are plugged into a backplane in the RTU cabinet. The RTU cabinet usually accommodates inside an environmental enclosure which protects it from extremes of temperature, humidity, etc. The following factors are the typical considerations while selecting the enclosures.

1. Circulating air fans and filters: This should be installed at the base of the RTU enclosure to avoid heat build-up. Hot spot areas on the electronic circuitry should be avoided by uniform air circulation. It is important to have a heat soak test too,

2. Hazardous areas: RTUs must be installed in explosion proof enclosures. In substations it is recommended to keep a safe distance from the circuit breakers,

3. Operating temperatures: Typical operating temperatures of RTUs are variables when the remote monitoring is located outside the building in a weather-proof enclosure. These temperature specifications can be relaxed if the RTU is situated inside a building, where the temperature variations are not as extreme (provided consideration is given to the situation, where there may be failure of the ventilators or air-conditioning systems),

4. Humidity: Typical humidity ranges are at the high humidity level that there is no possibility of condensation on the circuit boards or there may be contact corrosion or short-circuiting. Lacquering of the printed circuit boards may be an option in these cases. Be aware of the other extream, where low humidity air (5%) can generate static electricity on the circuit boards due to stray capacitance. CMOS based electronics is particularly susceptible to problems in these circumstances. Only screening and grounding the affected electronic areas can reduce static voltages. All maintenance personnel should wear a ground strap on the wrist to minimize the risk of creating and transferring static voltages, and

5. Electromagnetic Interference: Excessive electromagnetic interference (EMI) and radio frequency interference (RFI) is anticipated in the vicinity of the RTU, special screening and earthling should be used. Some manufacturers warn against using handheld transceivers in the

neighbourhood of their RTUs. Continuous vibration from vibrating plant and equipment can also have an unfavourable impact on an RTU, in some cases. Vibration shock mounts should be specified for such RTUs. Other areas which should be considered with RTUs are lightning (or protection from electrical surge).

2.2.4 RTU DESIGN STANDARDS

The RTUs available in the market are generally designed in accordance with applicable International Electro technical Commission (IEC), Institute of Electrical and Electronics Engineer (IEEE), American National Standards Institute (ANSI), and National Equipment Manufacturers association (NEMA) standards. For easy maintenance the architecture design generally support pluggable modules on backplane. The field wiring are terminated such that these are easily detachable from the I/O module.

2.2.5 SELECTION CRITERIA OF RTUs

The SCADA RTUs need to communicate with all on-site equipments and survive under the harsh conditions of an industrial environment. Hence the following points may be generally ensured while selecting the RTU.

1. Sufficient capacity to support the equipment at the site and the scope for the future expansion. Hence at every site, the RTU (Remote Telemetry Unit) that can support expected growth over a reasonable period of time, but within the budget may be selected,

2. Rugged construction and ability to withstand extremes of temperature and humidity. Depending upon the environmental factors where the RTU is intend to be kept, a rugged design which can keep the RTU system as the most reliable element,

3. Secure, redundant power supply. The SCADA system has to be up and working 24x7 without any excuses. RTU should support battery power ideally with redundant power inputs,

4. Redundant communication ports. Network connectivity is as important to SCADA operations as a power supply. A secondary serial port or internal modem may be kept in the RTU online to counter the LAN failure. Further RTU with multiple communication ports can easily support a LAN migration strategy,

5. Non-volatile memory (NVRAM) for storing software and/or firmware. NVRAM retains data even when power is lost. New firmware can be easily downloaded to NVRAM storage, often over LAN - so that capabilities of RTUs may be kept up to date without excessive site visits,

6. Intelligent control. Sophisticated SCADA remotes can control local systems by themselves according to programmed responses to sensor inputs. Though it is not necessary for every application, but it does come in handy for certain cases,

7. Real-time clock for accurate date/time stamping of reports, and

8. Watchdog timer to ensure that the RTU restarts after a power failure.

2.2.6 SECURING REMOTE TERMINAL UNITS

As RTU being one of the critical component of the DCS and ICS, physical-cyber security of the RTU is most important especially when deployed in critical infrastructure SCADA such as Power System SCADA (PSS), Oil and Gas Industry, etc. Today, the Electric utilities are most concerned of statutory security requirements of North American Electric Reliability Corporation Critical Infrastructure Protection (NERC CIP). The flexibility and range of RTU technologies necessitates a proper incorporation of its security and safety procedures.

Physical security of an RTU can be ensured by keeping restricted and authenticated access. Door opening of RTU enclosure can be properly monitored by a status signal, or by placing a surveillance camera which communicates the MCC in real-time RTU to alert system operators regarding the physical security breach of a remote RTU. One of the main standards that define procedures for implementing electronically secure Industrial Control Systems (ICS) is ISA/IEC-62443. This guidance applies to end-users, system integrators, security practitioners, and control systems manufacturers who design, implement, and manage the ICS. While dealing with Power System SCADA (PSS), most of the utilities are following the NERC CIP standards as they are exclusively developed for Bulk Electric System (BES).

NERC CIP defines a Critical Asset, as the facility, or system which, if destroyed or degraded, would affect the smooth and proper operation of the Bulk Electric System (BES). In other way, they are the programmable electronic devices and communication networks which comprise hardware, software, and the data. The Critical Cyber Assets (CCA) is defined as Cyber Assets, essential to the reliable operation of Critical Assets. Bulk Electric System (BES) are defined as, the electrical generation resources, transmission lines, sub stations, interconnections with neighbouring systems, and associated equipment, generally operated at voltages of 100 kV or higher. Obviously the RTUs installed at substations which deals with voltages more than 100kV or higher is a Critical Cyber Assets and need NERC CIP compliance and protection to ensure safety and security.

In order to frame the security policies and procedures, it is critical for utilities to understand the necessary RTU functionality and how it can be incorporated to achieve goals. Majority of the RTUs which are Critical Cyber Asset may not be kept in a completely protected area, and may be accessible to people who are not explicitly authorized to access them. This necessitates the key requirement of physical security of RTUs. Most RTU equipment includes, digital input points as a standard feature, and adding a door alarm is a simple way to meet NERC CIP requirements. Including the door alarm input as a sequence of events (SOE) point provides a means of meeting another aspect of the NERC CIP procedures for logging data and maintaining historical records.

Cyber security cannot be mentioned without encryption. Encrypted data transmission and reception makes RTUs more secure. It is obviously harder for an unauthorized user to manipulate the data sent and received. Encrypted data, however, has a disadvantage. Most notably, it will impact the scan rate of the device. Further encrypting and decrypting every message received and transmitted by the RTU amounts to a considerable increase in the processing time, which in turn affects the latency. This demands high computing power and processing capability of RTU CPU. Many RTUs being used in the field are already pushed to the upper limits without consideration for encryption. Another disadvantage is that the data is encrypted and the user will need special tools to view it, as it is often important to examine the data exchanged between devices to ensure they operate properly.

An important feature of RTU is its ability to provide logs of important events. Many of the alarms need to be logged for time of occurrence and the relevant information. Maintenance engineers who are making connections and configuration changes to the unit can be properly evaluated with this chain of events. Sequence of Event (SOE) logs, user logs and system logs can all be used to document events from the safe and secure operations of RTU.

The formation of procedures and plans, which need to be written with current and future capabilities in mind, is most important part of the SCADA security standards such as NERC CIP, IEC 62443, etc. Due to the wide technology range it will not be practical to replace or upgrade every device as quickly or easily. Securing an ICS or DCS is not a job for the faint hearts; rather it is a job that must be done with courage, intelligence, and imagination. A work which requires many hours of hard, thoughtful work for developing a security policy in tune with the utilities general safety and security policy. Today, it is considered that providing cyber security to critical infrastructure is no way inferior to the job of defending a nation by the army generals.

2.3   INTELLIGENT ELECTRONIC DEVICES (IEDs)

IEDs are one of the innovations to the power system which drastically changed the power monitoring, control and protection. Microprocessors based voltage regulators, protection relays, circuit breaker controllers, etc. with the capability of serial communication with other devices are referred as Intelligent Electronic Device (IED). The industry standard definition of an IED is, any device incorporating one or more microprocessors with the capability to transmit and receive data/control from or to an external device such as electronic multifunction transducers, numerical digital relays, bay controlling units, etc. Presently IEDs have been deployed extensively in power system automation, for monitoring, protecting, controlling, metering, and communicating. The flexibility in integration and interoperability of IEDs over RTUs is apparent in power industry automation. Today one cannot envisage power system automation without IEDs. Hence its evolution and functionality are briefly presented in the following sections.

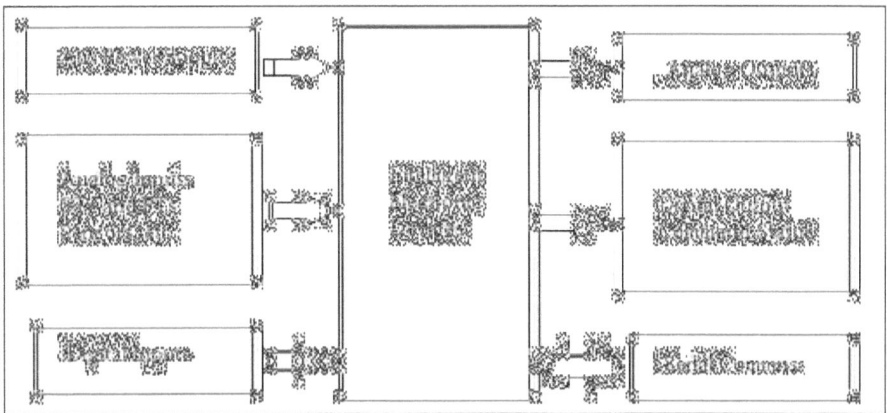

Figure 2.2 Block diagram of an IED

Figure 2.2 depicts the organizational block diagram of a typical intelligent electronic device. The modern IED architecture ensures that the device is multipurpose, modular in nature, flexible and adaptable, and has robust communication capabilities which include multiple selectable protocols, multi-drop facilities with multiple ports, and rapid response for real-time data. Current IEDs coming with tremendous computing power which is capable to carry out a variety of functions, various applications like protection and metering. IEDs have Sequence of Event (SOE) recording capability that can be very useful for post-event analysis, for fault waveform recording, and for power quality measurements. The advantage of this is, it eliminates the requirement of additional digital fault recorders and power quality monitors. IEDs can also accept and send out analog and digital signals with selectable settings, thus making the IEDs versatile. The details of the IED building blocks are briefly explained below.

Auxiliary power supply: Unlike older protection relays which may not need an auxiliary supply, IEDs always require an auxiliary power supply. Most IEDs accept an extended range of power supply, usually ranging from 14 - 150 V DC or 110 - 140 V AC.

Analog inputs: Protection relays are always provided with current transformer (CT) and potential transformer (PT) inputs. In addition, IEDs may be provided with sensor inputs. Hence it is important to specify the rated secondary current and frequency before ordering.

Digital inputs: Some IEDs require potential-free contacts for digital inputs (DI), while others recognize the positive power supply voltage (source) or negative power supply voltage (sink) as a logical '1'. Digital inputs may be commands or status information.

Analog outputs: Some IEDs are provided with transducer outputs. Mostly these outputs are programmable. These outputs can be active type or passive type outputs. The passive type requires external power supply.

Digital outputs: Digital outputs can be potential free normally open (NO), normally closed (NC) or solid state contacts. It is important to check the switching capability of the output contacts, as the differences can be significant. Digital outputs may be commands or status information.

IEDs have the real-time and rapid data exchange capabilities with multi ports. Many IEDs which are available in the market are capable of handling Analog Inputs, Discrete Outputs, Analog Outputs and Discrete Inputs. SoE recording, fault recording, metering and protection are the common features with all IEDs.

The IED integrates many single-function electromechanically relays, control switches, extensive wiring, and much more into a single unit. In addition the IED handles additional features like self and external circuit monitoring, real-time synchronization of the event monitoring, local and substation data access, programmable logic controller functionality, and an entire range of software tools for commissioning, testing, event reporting, and fault analysis.

2.3.1   IED Hardware and Software

The design of the IED architecture is always done in such a way that it ensures the programming, commissioning, and maintenance in a simple and convenient way. The hardware should be designed with the future adaptability requirement in mind, whereas the software structure should ensure the independent protection, control, metering, and communication functions. The hardware design of IED utilizes Plug-in type cards to a back plane which has great advantage in maintenance, as the replacement can be done easily without disconnecting the terminal wires and removing the IED from the panel.

The IED software is usually designed in such a way that the installation and commissioning engineer can easily evaluate the performance and configure the available functions independently. The required function can be easily selected and hide the functions which are not required. Each selected function is a self-determining independent system in the IED with dedicated logical inputs and outputs, setting, and event reporting features. Advanced IEDs have the capability of waveform capture and disturbance analysis. Metering and demand values recording, programmable logic capabilities are other features of the IEDs which eliminates an additional PLC usage.

2.3.2 IED COMMUNICATION MODULE

IED communication is the area of extreme significance as it is one of the unique features. Different manufacturers use different communication protocols, which provides flexibility and at the same time major benefits to the utility. IED which support different protocols for multi-port communication and different communication media and have flexible and open communication architecture is a general requirement today. In addition to the open communication protocol support, modern IEDs usually come with a serial electrical port, remote access port with modem, or optical port for fibre optic communication interface. This feature enables the IEDs to communicate directly to a DCU or HMI or other IEDs or Laptop for configuration and data downloading purposes.

To have the interoperability feature, open protocols are always preferred today. Modern IEDs are available in the market with communication modules which support a variety of protocols especially IEC 61850, DNP3, Modbus, etc. The advantage of these modules is that they can be replaced in the field in case of a change in communication requirement. Many IEDs are capable of multi-port communication and can communicate with substations and other IEDs at the same time through a modem to office or home or substation. As far as the physical layer is concerned, several serial communication ports are possible such as RS 232, RS 485, Ethernet (RJ45), optical, USB, etc. IEDs are mostly also provided with an RS 232 or USB port for local communication with a laptop or PC.

2.4   PROGRAMMABLE LOGIC CONTROLLER (PLC)

A Programmable Logic Controller (PLC) is a computer based solid state device that controls industrial equipment and processes. It was initially designed to perform the logic functions executed by relays, drum switches and mechanical timer/counters. Advanced PLC with Analog control capability are also available today, but RTU based SCADA systems are preferred by power utilities as the RTU can also function as Data Concentrating Unit (DCU) and offers more security especially when unguided remote communication is used. The advantage of a PLC over the RTUs which are available in the market can be used in a general purpose role and can easily be set up for a variety of different functions. The actual construction of a PLC can vary widely and are popular for the following reasons.

1. PLC offers more economical solution when compared to general purpose RTU solution,

2. The logic of the PLC can easily be modified to cope with new situations with improved requirements,

3. Ease of design and installation. Many hardware requirements can be suitably substituted with software in PLC. This makes the design and installation of SCADA systems easier,

4. If properly installed, PLCs are a far more reliable solution than a conventional hardwired relay solution or short run manufactured RTU,

5. PLCs allow for far more sophisticated control (mainly due to the software capability) than RTUs.

6. PLCs are very compactable. Hence occupy less space when compared to alternative solutions.

7. Trouble shooting is simple, easy and quick in diagnosing of hardware/firmware/software problems on the system

2.4.1   LADDER LOGIC

Ladder diagrams or ladder logic are, a type of electrical notation and symbology frequently used to illustrate how electromechanical switches and relays are interconnected. It is a rule-based programming language rather than a procedure language, which creates and represents a program through ladder diagrams. It is mainly used in developing programs or software for PLCs.

Ladder logic is widely used in industrial settings for programming PLCs where sequential control of manufacturing processes and operations is required. The programming language is quite useful for programming simple yet critical systems or for reworking old hard-wired systems into newer programmable ones. This programming language is also used heavily in highly sophisticated automation systems such as electronics and car factories.

The idea behind ladder logic is that even personnel without programming backgrounds can quickly program, since it makes use of conventional and familiar engineering symbols for programming. But this advantage is quickly negated since manufacturers of PLCs often also provide ladder logic programming systems with their products, which sometimes do not use the same symbols and conventions as those made for other models of PLCs from other manufacturers; in fact, the programming system is usually meant only for specific models, so the programs cannot be ported easily to other PLC models or must be outright rewritten.

2.5   DATA CONCENTRATORS AND MERGING UNITS

Another two important building blocks that are very useful in modular design of DCS are Data Concentrators and Merging Units. A brief description is given below.

2.5.1   DATA CONCENTRATION UNITS (DCU)

In certain situations the DCS design, especially in PSS, is an intimidating task because of the extremely large number of input and output data. The variety of large number of field devices and IEDs with different protocols makes the DCS design very complex. Considering the fact that most substations are automated and intended to operate unmanned, these data has to be transmitted to MCC for control and analysis. The RTU in a substation can serve as a data concentrator by gathering or concentrating the data from the field devices. It utilizes a data concentrating device that has lots of I/O ports on it. When connected to all the IEDs and field devices which can send their details as data in a star topology, this devices are polled to gather the data and transmits it to a remote master and/or client server stations. I/O ports on modern data concentrators available with Ethernet ports such as fibre ports and RJ45 ports suitable for CAT5 cable. These ports can enable a TCP/IP communication and establish a LAN or WAN connectivity. Data concentrators offer several additional capabilities. One of them is to serve as a gateway device. That is, any connection to a WAN system is made at only one point in the substation network and is using this device alone. Modern gateway devices, suitable in DCS environment have the following capabilities.

1. It has the capability to set up a firewall, with SCADA specific IDS loaded to keep the intruders away,

2. It has the capability to provide a secure connection for secure data transmission. Many RTUs/DCUs have the NERC CIP compliance. Remote connection with DCU can be established via the internet using VPN, leased telephone lines, OFC, or unguided medias,

3. Modern RTUs support a number of protocols. They also have protocol conversion capabilities. This helps the field devices which do not support modern SCADA protocols such as DNP3 or IEC 61850, can also be connected and communicated with the RTU without replacing the legacy IEDs, and

4. Many RTUs have the capability of providing a web based HMI. In certain cases, an HMI has to be kept at remote locations with limited capability or use it as a Local Data Monitoring System (LDMS). In such situations, accommodating devices like a display monitor, keyboard, and mouse, etc. must be connected to the RTU or LDMS in a secure manner.

If the substation is small, then all station IEDs and field devices can be connected directly to the data concentrator. No Ethernet switches are required and additional I/O ports on this device can be installed if necessary.

2.5.2   Merging Unit (MU)

Merging Units (MU) is also an important element in the design of modern DCS, especially in power industry when dealing with Substation Automation. Integration with the Merging Units take the local area network to another level called the process bus, right into the field. The MU provides the suitable interface for the implementation of the process bus concept in modern substation automation systems. Today the IEC 61850 is becoming the standard for the communication architecture for SCADA networks of the substations. The basic function of the MU in automation systems is to convert the analogue values of current and voltages measured at the process level to digital format and transmit same to the microprocessor relays. In substations, these multichannel digital signals output by electronic current and electronic voltage transformers are collected synchronously and transmitted with the protocol of IEC 61850 to protective, measurement and control devices. The use of merging units eliminate several multiple wire connections running from the switchyard to the microprocessor relays located in the control room. The conventional merging unit model collects current and voltage signals from various current and voltage transformers in the switchyard, converts them into digital form and sends the digital equivalents to the microprocessor relays via a single fibre optic cable known as the process bus. Today modern merging units are available with additional features of in-built overcurrent protection and bay control functions. These new models are intended to provide local over-current protection and bay control for all equipment in the bay being monitored by a particular merging unit.

2.6   MASTER CONTROL CENTRES (MCC)

Figure 2.3 shows a large master control centre with a dual redundant LAN, with all the components, designed with full redundant hardware and software features to achieve High Availability (HA). The system redundancy and communication channel redundancy must be ensured from remote location to the servers of the master station. The master station must be fault tolerant and fail safe in every respect, so that any natural calamity affecting one station will not cause a problem with the functioning of other stations and the system can be monitored and controlled effectively. Uninterrupted power supplies (UPS) monitoring systems are also very important and must be ensured as most of the process /plants are non- stoppable and cannot be interrupted. The various SCADA components in the MCC are elaborated below.

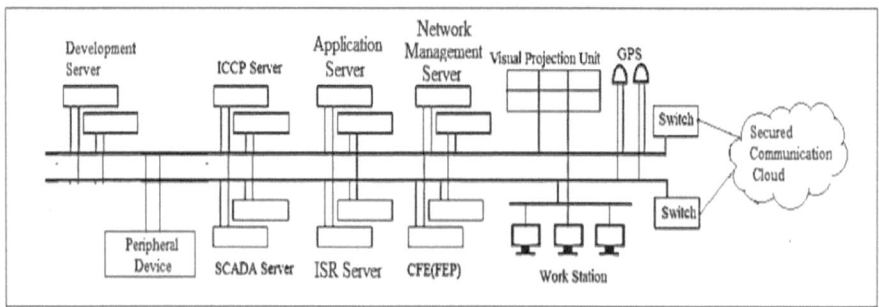

Figure 2.3   Block Diagram of a Typical Master Control Centre

2.6.1   SYSTEM SCADA SOFTWARE

These are the data acquisition and control, database design and development, reporting and accounting, and the HMI software. It performs all the basic functionalities which are required in common to all SCADA application. As far as power system SCADA is concerned, all SCADA functions are considered as critical functions.

2.6.2   MASTER STATION HARDWARE

The main hardware in a master station are the computer, server systems, display units, printers, plotters, routers, and security and protecting devices such as firewalls in addition to uninterrupted power supply ensuring the power quality. Many servers are used for executing the different tasks to be performed. Hence selection of servers is most important and selection must be based on the requirements of the master station.

2.6.3   SERVERS IN THE MASTER STATION

As mentioned, different dedicated redundant, high availability servers are deployed in the SCADA master control station to execute specific tasks.

The servers are connected through a high speed dual redundant LAN. The client server architecture ensures that the data is accessed by another server whenever it is required and properly authenticated. The dedicated server systems should have special capabilities and features depending upon the requirements of the application for which it is deployed. Some of these features are CPUs with high computing and fast processing power, high-performance RAMs, redundant and uninterrupted, quality assured power supplies, high end routers and switches for network connections, and high performance Firewalls.

The computer server systems available in a SCADA master station are as follows:

1. SCADA server,
2. Application server,
3. Information Storage and Retrieval (ISR) Server,
4. Development server,
5. Network Management Server (NMS),
6. Video Projection System (VPS),
7. Communication Front End Server (CFE),
8. Inter Control Centre Communications Protocol (ICCP) server, and
9. Dispatcher Training Simulator (DTS) server.

The main functions of each server are described below.

2.6.4 SCADA SERVER

The SCADA server is assigned with the data acquisition of all inputs and present it to other servers for displaying, further processing and decision making. It also accomplishes the control command execution as per the directions issued from the SCADA control station. In small SCADA systems the SCADA server directly gathers the data from the field devices and issue control commands directly to the devices. But when the SCADA system is large SCADA server communicates through the CFE server.

2.6.5 APPLICATION SERVER

Depending upon the nature of process/plant to be controlled, the SCADA application varies. It can be EMS, GMS or DMS as far as power system is concerned. The application software modules are hosted by the application server required for the specific SCADA system.

PSS Distribution Management may have voltage reduction, load management, power factor control, two-way distribution communications, short-term load forecasting, fault identification, fault isolation, service restoration, interface to intelligent electronic devices {IEDs), three-phase unbalanced operator power flow, interface to/integration with automated mapping/facilities management (AM/FM), interface to customer information system (CIS), trouble call/outage management, and so on. The transmission SCADA may have the Energy Management Systems (EMS) package, which includes network configuration/topology processor, state estimation, contingency analysis, three-phase balanced operator power flow, optimal power flow, etc.   For the Generation Management PSS, the application software may include Automatic Generation Control (AGC), economic load dispatch, unit commitment, short-term load forecasting, etc. In case of controlling the refinery plant, necessary process monitoring and control software is loaded in the SCADA server. Same is the case with Water supply and management system.

2.6.6 INFORMATION STORAGE AND RETRIEVAL (ISR) SERVER

An information storage and retrieval is historian server which has the following functionalities.

1. Supports the reporting accounting activities,
2. Archiving of data for the system,
3. Real-time data snapshot, and
4. Historical information recording, retrieval, and report generation.

2.6.7 DEVELOPMENT SERVER

All preliminary engineering and commissioning activities of the Power System SCADA (PSS) is done with the development server. Further the changes or developments thereafter have to be handled to keep the system real-time. The various programs such as application software development, display development, and database generation are developed also by the development server.

2.6.8 NETWORK MANAGEMENT SERVER (NMS)

NMS may be used to monitor both software and hardware components in a network. It is an application or set of applications that helps the network administrator to manage network's independent components inside a larger network. It usually records data from a network's remote points to carry out central reporting to a system administrator. Modern control centres have many digital devices connected via the local network of the master station. In fact a network management system manages the following functions in a SCADA control centre.

1. Network device discovery,
2. Network device monitoring,
3. Network performance analysis,
4. Network device management, and
5. Intelligent notifications or customizable alerts.

2.6.9 VIDEO PROJECTION SYSTEM (VPS)

In large master control stations, especially in power industries where SCADA is employed, the distribution or transmission network is displayed using a video projection system. Master stations must be equipped with high-tech display systems that can display the area of control in a diverse manner as per the requirement of the operators. This display can be GIS information, a Single Line Diagram of the network of the region of interest and a separate video projection system handles this function. Supporting software helps in tracing the distribution and transmission network.

2.6.10 COMMUNICATION FRONT END (CFE)

Communication front end (CFE) interfaces the host computer to a network or peripheral devices. CFE is used to unburden the host computer input and output communications such as managing the peripheral devices, transmitting and receiving message packet assembly and disassembly, and error detection and error collection. This CFE, often referred to as the FEP (front-end processor) must be a high end machine with high computing capabilities so that the communication with a large number of the peripheral devices could be carried in real-time. FEP communicates with the host computer using a high-speed parallel interface. Obviously these computers are costly. FEP/CFE is synonymous with the communication controller. In power system SCADA, CFE has the role of communicating with all RTUs, other field devices which mostly connected through FRTUs, RVDU (Remote Video Display Units), etc.

2.6.11   ICCP Server

ICCP server is an important one in Power system SCADA. This inter-control centre protocol server supports bulk data transmission between the master stations depending on the hierarchy. Typically, the ICCP server at a Regional Load Dispatch Centre (RLDC) exchange data between RLDC and National Load Dispatch Centre (NLDC) as well as with the State Load Dispatch Centre (SLDC) and sub-LDC if required but maintaining the hierarchy.

2.6.12   Dispatch Training Simulator (DTS) Server

DTS is also an important server in Power System SCADA. A Dispatcher Training Simulator (DTS), also known as an operator training simulator (OTS), is a computer-based training system for operators/ dispatchers of electrical power grids, is generally available at a large master station. It is used to train the dispatchers who manage the system. The DTS generally provides the power system model, hydro system model, control centre model, and instructor functions. It performs this role by simulating the behaviour of the electrical network forming the power system under various operating conditions, and its response to actions by the dispatchers. Trainees may therefore develop their skills from exposure not only to routine operations but also to adverse operational situations without compromising the security of supply on a real transmission system. Modern DTS combines or simulates the elements such as Energy Management System (EMS), Distribution Management Systems, SCADA system, load-flow study, and facilities for modelling and optimizing the economic dispatch of generating units.

2.7   GLOBAL POSITIONING SYSTEMS (GPS)-RELEVANCE TO SCADA

A space based radio navigation system which can provide time information and relocation to a GPS receiver which is placed on or near the Earth surface which has four or more unhindered Line of Sight (LoS) to four or more GPS satellites. There are 14 to 31 geostationary satellites in the mid orbit of the earth which are owned by United States and operated by US Air Force.

Presently time synchronization for all Industrial SCADA components especially in PSS and Smart Grid including the master control station are achieved with the GPS clocks. The GPS system does not require the user to transmit any data, rather it continuously transmit signals to the earth which contain information about the time of transmission and the satellite locations at the time of transmission. On receiving these signals, GPS receivers compute the distance to the satellite based certain equations, the location of the receiver and the time. GPS time is theoretically accurate to about 14 nanoseconds. However, most receivers lose accuracy in the interpretation of the signals and are only accurate to 100 nanoseconds. In PSS and Smart Grid, GPS plays a very vital role, as all the time stamped events, SoE, IED and PMU measurements, etc. critically needs time synchronization.

2.8   HUMAN MACHINE INTERFACE (HMI)

Human Machine Interface (HMI) or user interface (UI) is a combination of software and hardware that allows the interaction between the humans and the system (machine) in a SCADA environment. The main objective of this interaction is effective operation and control of the system being monitored. This aids the operator to make operational decisions and manually override automatic control operations in the event of an emergency. In local and small SCADA control centres, HMI allows a control engineer or operator to configure set points or control algorithms and parameters in the controller. In medium and large SCADA control centres, it also provides process status information, historical information, reports, and other information to operators, administrators, managers, business partners, and other authorized users.

Today the devices and instruments that an operator used in PSS have changed significantly from manual to computer-based devices. The latest expensive hardware, processors with high computing capability, dedicated software, mimic diagrams, and communication protocols with security features especially with end node security have made the system very efficient, compact and human friendly. The location, platform, and interface of HMI may vary a great deal. Further present day HMI could be a dedicated platform in the control centre, a laptop on a wireless LAN, or a browser on any system connected to the Internet.

2.9   HMI BUILDING BLOCKS

In a SCADA system, the HMI components include operator console, operator dialogue, mimic diagram and peripheral devices, etc. are briefly explained below.

2.9.1   OPERATOR CONSOLE

The console where the operator monitors and controls the system is of utmost importance and includes the visual display units, alphanumeric keyboard, cursor, communication facilities, etc. The Visual Display Unit (VDU) includes UI devices like the multiple color monitors (CRT, LCD, LED devices minimum size), with glare-reduction features (antiglare screen coatings) and should provide a display of multiple viewports (windows) on each monitor. The cursor control could be mouse, trackball, or the latest touch-screen facility. A keyboard and cursor pointing device are shared among all monitors at each console and the cursor moves across all screens without switching by user. Generally for power system SCADA each operator has three to four monitors for proper planning and multiple views and the displays should have full graphics capability with zoom facility. Audible alarms are also a prominent feature of the operator console where the operator is informed of the severity of an event in the system. The design of the operator console infrastructure including the table and chair for the operator is important and should follow ergonomics principles to make the operator comfortable during the duty period.

2.9.2 OPERATOR DIALOGUE

Operator dialogue box is a box that pops up to enable communication between the computer and the operator. Dialogue boxes may ask you questions or give you information. The operator dialogue and commands should be simple and easy to remember. Certain boxes may ask for an action relating to the application which is being used. Function keys of the keyboard can be programmed to incorporate major actions so that the operator can give the commands effortlessly rather than typing long dialogues.

2.9.3 MIMIC DIAGRAM

The mimic diagram is an essential part of any control centre or large master station where the operator and the personnel in charge get an overall view of the plant/process under control. This includes LCD/LED large-screen display with full SCADA operability with multiple screens possible. Some control centres have mosaic map board with dynamic or static tile map board and dynamic map board lamps updated by SCADA. Present trend is to use multiple video projection cubes as the dynamic map boards. The major benefit is that when the HMI is updated with system changes, the map board is also automatically updated, since the HMI drives the map board directly.

2.9.4 PERIPHERAL DEVICES

Generally three printers are required in Master Control Centres. One of them is used to print alarms and SoEs. Another printer usually a colour printer, which is used for capturing screen shots. A black and white laser printer is used to print reports.

2.9.5 HMI SOFTWARE FUNCTIONALITIES

Selecting HMI software typically starts with an analysis of product specifications and features. The key considerations can include the system architecture, performance requirements, integration and cost of procurement and operations which are described below.

1. The HMI console should have Role Based Access Control of security to protect unauthorized access to the system. As this HMI is one of the end node of the SCADA system, specific user identification (IDs) and password must be used.

2. The display of power system to control may be in an effective way. There should be a provision to display all the information about the power system interconnection and the parameters of interest in the HMI, such as voltage, current, frequency, and power flow. This must be in an user friendly manner so that even a new operator could monitor and analyze the events and take corrective and control action, if needed.

3. HMI software should have the capability of preparing Logs and Reports, and Calculated Values. These reports are required many times to present to various system hierarchies and to different departments of the utility. Today more sophisticated HMI software is structured around mobile, portable platforms. This presents a cost-saving value as the operating systems are distributed on machine-level embedded HMI, solid-state open HMI machines, distributed HMI servers and portable HMI devices.

2.9.6 SITUATIONAL AWARENESS AND ALARM HANDLING

Situational Awareness: Situational awareness is be aware of what is happening around the operator in the control room, so that he will be able to decide how to react to the events. The operator in a control room should observe the environment, comprehend the situation, and take decisions and actions accordingly. Operators dealing with critical functions play an integral role in the safe and efficient operation of the system. Situational awareness is important when the information flow is fast and the error in judgment will lead to major consequences. In power systems, when a blackout suddenly occurs an appropriate action should be taken on time, which needs a rapid decision making and execution. Compared to other SCADA systems, operating a Power System SCADA (PSS) is more complex and mentally demanding, and thus PSS operators are more vulnerable to human errors.

HR experts consider situational awareness, as very important one, being the perception of elements in the environment within a volume of time and space, the comprehension of their meaning, and the projection of their status in the near future. Special considerations to mitigate operator errors should be taken when designing an operator-assistance system for PSS. A model developed by psychologists includes three levels viz.

1. *Perception level:* In this level the person has to perceive the status, attributes, and dynamics of the variables in the environment,

2. *Comprehension level:* In this level the data from level one have to be synthesized using interpretation, pattern recognition, and evaluation skills, and

3. *Projection level:* Here the person can extrapolate the information from the lower levels and arrive at an action plan.

Operators may make errors when they are not totally aware of the actual situation. So one of the important objectives is to equip the Master Control Centre with appropriate visual aids to meet the required perception level. Provide enough data processing and display systems for the comprehension level and finally enable the operator to take a decision and execute it. Today, SCADA software visualization and display have undergone a major transformation with new devices and tools available for increasing the perception and comprehension levels of the operators, to equip them to make better decisions at the projection level.

Alarm Handling: Once the data are communicated to the control centre, they are first processed before presentation to the operator. The processed data are then compared with the predefined values and in case of any deviation from the nominal value, an alarm is generated. The operator acknowledges the alarm and appropriate remedial action is taken. Thus, important functions of the control centre allow it to generate, annunciate, and manipulate the process and system alarms. The generation and display of the alarm and the limits are important functions of the control centre, and this information needs to be communicated to the interconnected system in case of emergency, limit violations, and malfunctions. Sometimes, the operator is confused by the series of alarms triggered by a single event, as many quantities change and unnecessary alarms of all kinds are directed to the operator. Hence, alarm filtering is important and is explained in the following section.

2.9.7 Intelligent Alarm Filtering

Alarm handling and processing technology ensures that the SCADA operators receive only those alerts which are most relevant to events that must be addressed immediately. The details of less critical secondary warnings are sent to databases and possibly printed for later review. With only the most important distribution system alarms presented in a prioritized fashion, SCADA operators can assess problems more easily and make better decisions to prevent a crisis.

In a plant or process which is controlled by SCADA system, if the number of alarms annunciated are too large so that it can't be managed by the human operator leading to upset the plant or process operations is generally refereed as alarm flooding. This demands an effective alarm management technique to ensure that the alarms are presented to the operator at a rate which can be assimilated by the human operator.

In PSS especially in SCADA/DMS alarms are typically triggered by faults and the events surrounding them, which occur continuously during routine operations. When a breaker on a substation feeder trips due to a transient fault, for instance, up to seven alarms may be triggered: one for the breaker trip and three each when voltages and currents on all three phases hit zero. The dispatcher needs only the breaker trip alarm and does not need any alarm information if the breaker is automatically reclosed after a transient fault since the situation resolves itself.

Today, many industrial SCADA vendors prioritise the alarms as primary and secondary alarms depending on the urgency of action by the operators. Primary alarms require urgent operator action while secondary alarms need no operator action. Certain SCADA vendors developed alarm filtering techniques, so that alarms can be configured during SCADA implementation or activated during alarm flooding.

2.9.8 Necessities and Requirements of Operators

The SCADA master control centre is the place from where the operator monitors the plant or process operations and issues necessary control commands. Operator on duty in industrial SCADA has to spend long hours with a good presence of mind to monitor and control the plant or process operations. An ergonomic and aesthetic design of the control centre is a must to ease the burden of the operator and to avoid stress in vital body parts. An appealing ambience has to be created and the operator console design meets the standards so that, the operator is comfortable to perform his/her responsibilities with concentration and stress free. Some of the needs and requirements of the operator in the control room, to meet the functional objectives, are summarised below.

1. A pleasant ambience in the control room must be provided so that the operator working for long durations gets facilities, such as entertainment, exercise, refreshments, time for rest, etc,

2. The operator in the SCADA control centre has to handle many I/O devices through different consoles. Hence the I/O devices must be arranged within reach, and the operator may be able to move from one device to another very effortlessly, if needed,

3. Appropriate safety slogans may be displayed in the control room for reminding the operators for carrying out safe, secure and error free operations to avoid fatal and non-fatal accidents,

4. User-friendly operation and control system must be provided for smooth and error free operation,

5. Comfortable access to the control devices to the operator must be ensured and the alarms and indicating displays must be easy to locate,

6. The Video Projection System (VPS) which display the mimic and the other related information must be easily readable and also provide accurate and proper information to the operator,

7. The operator console, the input devices, the display units, etc. must be arranged in such a manner to provide comfortable eye-hand coordination approach for handling the equipment, and

8. In order to improve the working efficiency of the operator and reduce tiredness, provide comfort in the use of devices, stability and reliability of the equipment etc.

SUMMARY

This chapter begins with a comprehensive description of Remote Terminal Units, Programmable Logic Controller (PLC) and the different basic components of SCADA such as Intelligent Electronic Devices, Data Concentrator Units, Merging Units, Human Machine Interface, etc. The brief introduction of Data Concentrators and Merging Units are presented in such a manner that how digital substations can be designed. This chapter then gives an introduction of architecture of SCADA Master Stations and its hardware and software components. It concludes with a brief description of Geographical Positioning System (GPS), Situational Awareness and Alarm Processing.

Chapter THREE

SCADA ARCHITECTURE

3.1   INTRODUCTION

Various SCADA architectures are adapted today depending on the requirements. It starts with the basic single channel Data Acquisition and Control architecture to the complex multi-channel, real-time, fault tolerant, fail safe system which utilizes many secure communication methodologies with proper End Node Security (ENS). The following sections elaborate the various SCADA architectures with emphasis on Distributed SCADA.

3.2   COMMUNICATION ARCHITECTURE

There are three physical communication architectures which are generally popular and deployed in SCADA systems. In certain cases they are deployed in a combined mode. They are:

1. Point to point,
2. Point to multistations, and
3. Relay Stations.

3.2.1   POINT-TO-POINT BETWEEN TWO STATIONS

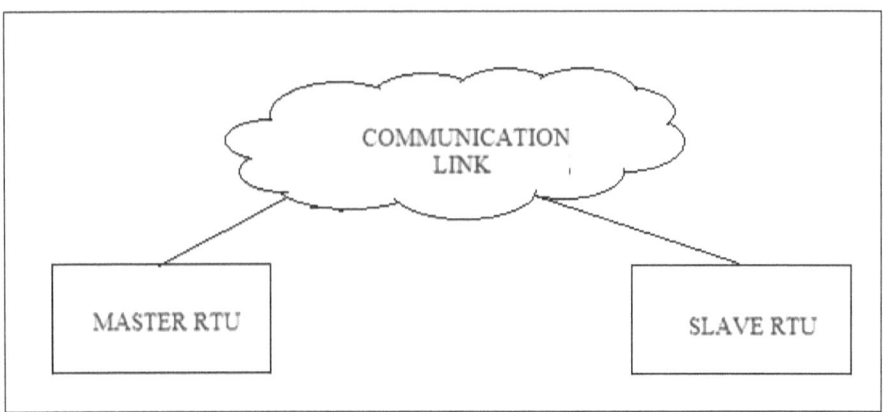

Figure 3.1   Point to Point Master-Slave Communication

This is the simplest configuration where data is exchanged between two stations. One station can be setup as the master and one as the slave as shown in Figure 3.1. It is possible for both the stations to communicate in full duplex mode (transmitting and receiving on two separate frequencies) or simplex with only one frequency.

3.2.2   MULTIPOINT OR MULTIPLE STATIONS

In this configuration, there is generally one master and multiple slaves which is shown in Figure 3.2. Generally data points are efficiently passed between the master and each of the slaves. If two slaves need to transfer data between each other they would do so through the master who would act as arbitrator or moderator. Alternatively, it is possible for all the stations to act in a peer-to-peer communications manner with each other. This is a more complex arrangement requiring sophisticated protocols to handle collisions between two different stations wanting to transmit at the same time.

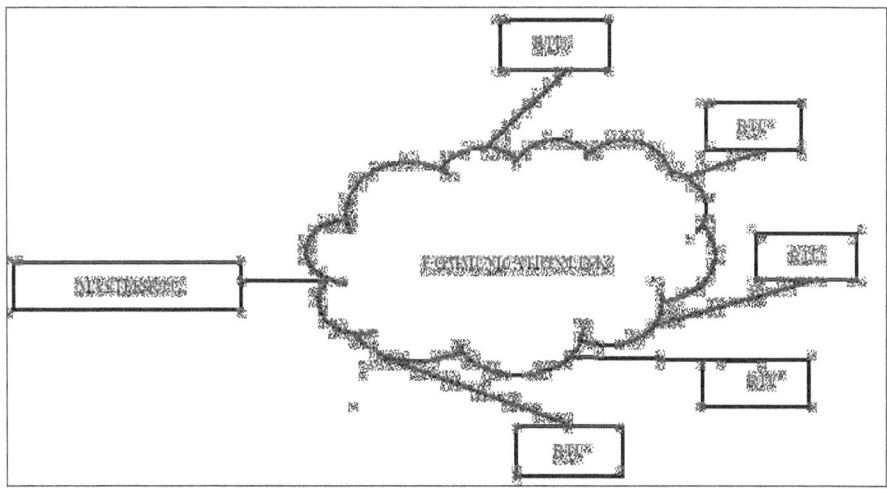

Figure 3.2   Master RTU communicating with multiple station RTUs

Another possibility is the store and forward relay operation. This can be a component of other approaches discussed above where, one station retransmits messages onto another station out of the range of the first station which is shown in Figure 3.3.

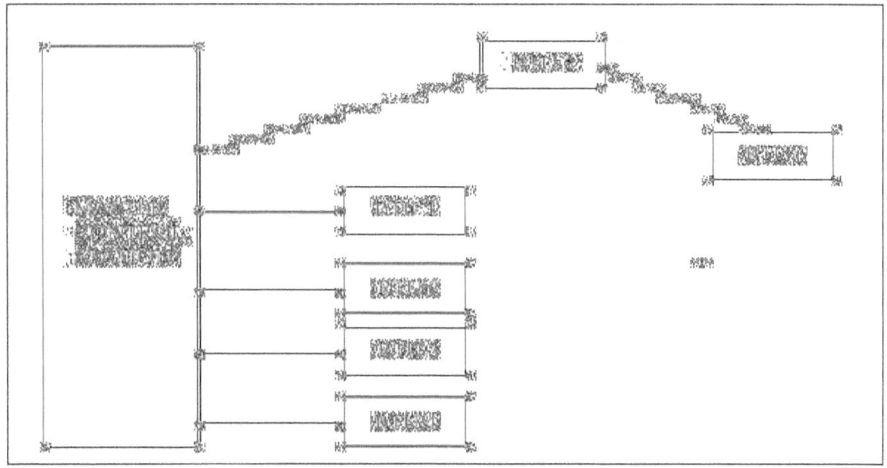

Figure 3.3   Store and forward station

3.2.3   TALK THROUGH REPEATERS

This is the generally preferred way of increasing the range of radio systems. This retransmits a radio signal received simultaneously on another frequency. It is normally situated on a geographically high point.

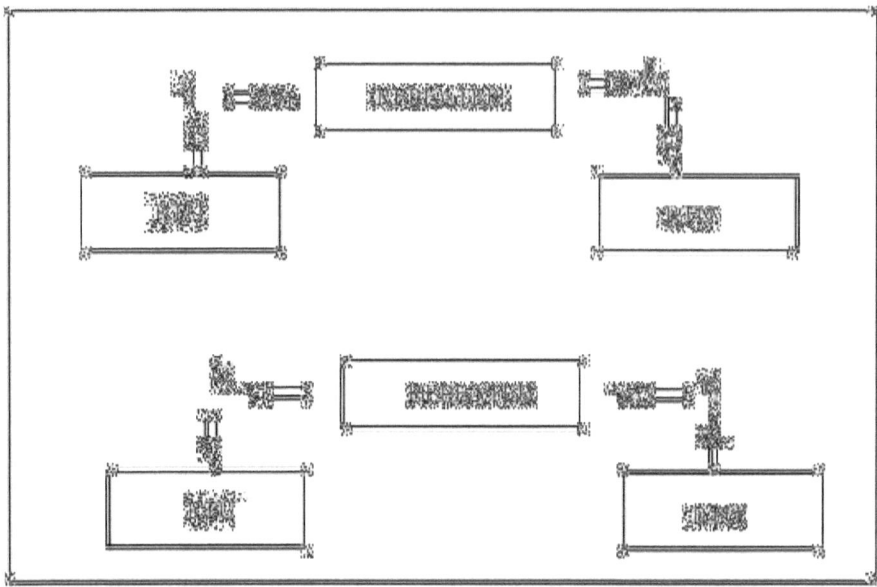

Figure 3.4   Talk through repeaters

The repeater receives on one frequency and retransmits on another frequency simultaneously. This means that all the stations repeating the signal must receive and transmit on the opposite frequencies. It is important that all stations communicate through the talk through repeater. It must be a common link for all stations and thus have a radio mast, high enough to access all RTU sites. It is a strategic link in the communication system; failure would wreak havoc with the entire system. The antenna must receive on one frequency and transmit on a different frequency which is shown in Figure 3.4. This means that the system must be specifically designed for this application with special filters attached to the antennas. There is still a slight time delay in the transmission of data with a repeater. The protocol must be designed with this in mind with sufficient lead-time for the repeater's receiver and transmitter to commence operation.

3.3   COMMUNICATION PHILOSOPHIES

There are two main communication philosophies in practice. These are polled (or master slave) and carrier sense multiple access/collision detection (CSMA/CD). The one of the notable methods accepted for reducing the amount of data that needs to be transferred from one point to another is to use exception reporting.

3.3.1   POLLED OR MASTER SLAVE

This is one of the simplest master- slave point to point or point to multipoint configuration where the master polls, the slave in predetermined regular intervals and gather the data which is shown in Figure 3.5. Here the master is in total control of the monitoring, decision making and control of the process. The slaves never initiate the data exchange rather, wait for the master's request and respond. Essentially it is a half-duplex communication. In case the slave fails to respond timely, then the master retries typically two or three times. If the slave still fails to respond, the master then moves to the next slave in the sequence with a remark that the particular slave is inactive or faulty and requires attention for rectification.

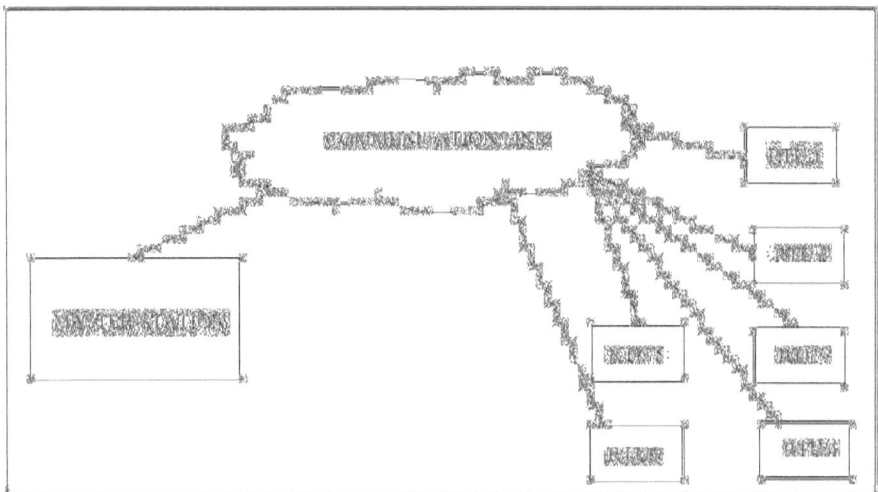

Figure 3.5   Illustration of master-slave polling with RTUs

The advantages of this approach are

1. Software development is fairly easy and can be made reliable due to the simplicity of the philosophy,

2. Link failure between the master and a slave node is detected fairly quickly,

3. No collisions can occur on the network, hence the data throughput is predictable and constant, and

4. For heavily loaded systems, each node has constant and bulk data transfer requirements giving a predictable and efficient system.

However it has the certain disadvantages which are described below.

1. Variations in the data transfer requirements of each slave cannot be handled,

2. Interrupt type requests from a slave cannot be entertained as the master may be either attending or processing some other slave request or data,

3. Systems, which are lightly loaded with minimum data changes from a slave, are quite inefficient and unnecessarily slow,

4. Communication between the slaves can be achieved only through the master which adds complexity.

Two applications of the polled or master slave, approach are given in the following two implementations. This is possibly the most commonly used technique and is illustrated in the Figure 3.6 below. The master station has a polling list with details of priority and sequence. Based on this information it prepares a polling cycle as shown in Table 3.1.

Table 3.1   Polling table with the master station

RTU 1
RTU2
RTU 5
RTU 6
RTU 3
RTU 4

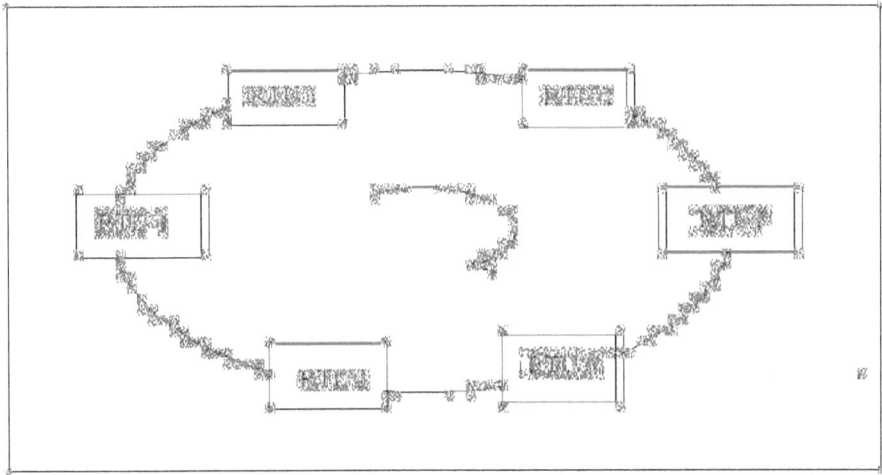

Figure 3.6   Polling cycle as per the polling table

In the scheme of polling, in certain situations, the polling may be modified as,

1. if there is no response from a given RTU during a poll, a timeout timer has to be set and three retries (in total) initiated before flagging this station as inactive, and

2. if an RTU is to be treated as a priority station it will be polled at a greater rate than a normal priority station. It is important not to put too many RTUs on the priority list, otherwise the differentiation between high and normal priority becomes meaningless.

3.3.2   CSMA/CD SYSTEM (PEER-TO-PEER)

3.3.2.1   RTU to RTU communication

In certain situations, especially in DCS an RTU in the SCADA system may need to communicate with another RTU. One of the solutions is, while responding to the master station to the poll, a message carrying a request with the destination address of the RTU can be added. The master station will then examine the destination address field of the message received from the RTU and retransmit onto the appropriate remote station. The only attempt to avoid collisions is to, listen to the medium before transmitting. If a collision occurs, the RTU wait for a random time period, and retransmit the data avoiding the collision. In this style of operation, it is possible for two nodes to try and transmit at the same time, with a resultant collision. In order to minimize the chance of a collision, the source node first listens for a carrier signal before commencing transmission. Unfortunately this does not always work where certain stations which cannot hear each other to try and transmit back to the station simultaneously.

3.3.2.2   Exception reporting (or event reporting)

On many occasions, the status and the RTU may be the same but the polling mechanism gathers the data from the RTU. This unnecessary transfer of data can be minimized or virtually eliminated with a technique called exception *reporting*. This approach is popular with the CSMA/CD philosophy but it could also offer a solution for the polled approach where there is a considerable amount of data to transfer from each slave.

In exception reporting, the remote station reporting devices such as RTUs monitor itself to identify a change of state or data. If there is a change of state, the remote station writes a block of data to the master station when the master station polls the remote. Typical reasons for using polled report by exception include:

1. The polling or scanning is performed with a low data rate due the communication channel constraints,

2. There is substantial data being monitored at the remote stations, and

3. The number of remote devices connected to master station is reasonably high.

Each analog or digital point that reports back to the central master station has a set of exception reporting parameters associated with it. The type of exception reporting depends on the particular environment but could be:

1. High and low alarm limits of analog value

2. Percent of change in the full span of the analog signal

3. Minimum and maximum reporting time intervals

The main advantages of this approach are quite clearly to minimize unnecessary (repetitive) traffic from the communications system.

3.3.2.3 Polling plus CSMA/CD with exception reporting

A practical method to combine all the approaches discussed earlier is to use the concept of a slot time and exception reporting. Here each slave station is assigned a specific time slot comprising the following sub-slot assuming that there is no requirement for communication between the slaves.

1. A slave transmitting to a master and
2. A master transmitting to a slave.

A slot time is calculated as the sum of the maximums of modem up time, plus radio transmit time, plus time for protocol message, plus muting time of transmitter. The master commences operations by polling each slave in turn. Each slave will be in synchronize with the polling and respond to the master with exception reporting if there is a status change, else respond with passive acknowledgement. As a result, the master move on to poll the next slave by overriding the remaining sub-slots. Otherwise it will complete the data exchange and then move to hear from the next slave. The master thus completes the poll cycle.

The previous and present chapter elaborated the building blocks of SCADA systems starting from the RTU, IEDs, communication systems, master stations and the HMI. Utilities have a variety of options available to mix and match the elements to building a cost-effective, efficient, and operator-friendly SCADA system as per their requirements.

Automation of the power systems started as early as the beginning of the twentieth century, and substations and control centers operate at various stages of automation all over the world. There are legacy systems with RTUs, hardwired communication from the field to the RTU, and traditional software functionalities in the control room, and it is not often financially viable to dismantle everything and purchase a completely new automation system.

Hybrid systems are a viable option, where any automation expansion project can be implemented with new devices, like IEDs, data concentrators, and merging units. The new system will coexist with the legacy RTU-based systems and the data integration and if necessary protocol conversion issues will have to be handled while commissioning the project.

If a utility decides to purchase a completely modern system, the latest building block of the SCADA system, viz, IEDs, merging units, and fiber optic communication facility with brand new HMI with situational awareness and analysis tools, it can be implemented.

3.4   SYSTEM RELIABILITY AND AVAILABILITY

Real-time system operation demands high level of availability and reliability to ensure error free operations. The real-time process control systems directly control the process, and any system catastrophe or erroneous operation may lead to process damage and may affect the safety of operations. Thus these applications necessitate a high level of system reliability without compromise. Numerically system reliability is defined as the probability that the system will not fail under specified conditions.

Standards and guidelines are available for the development of safe and secure critical systems. Among these standards, the most pertinent is for power system automation and is the NERC CIP standard. These development standards may be used while designing a power system SCADA. In keeping with these guidelines, an analysis must be performed as the early stage of system design in order to assign a system integrity level which allows the utility to define the accepted failure rate of system under consideration. Before discussing the fail safe system, it is better to have a clue of the two classification of failures namely Common cause failures and Common mode failure.

Common Cause Failure (CCF):   A Common cause failure occurs when two or more items fall within a specified time such that the success of the system mission would be uncertain. Item failures result from a single cause and mechanism.

Common Mode Failure (CMF):   A Common-Mode Failure is the result of an event(s) which because of dependencies, causes a coincidence of failure states of components in two or more separate channels of a redundancy system, leading to the defined systems failing to perform its intended function.

3.4.1   FAIL SAFE SYSTEM

A fail safe system describes a feature or a device which in the event of failure, responds in a way that will cause no harm or minimum harm to other devices or danger to personnel. Fail safe systems are used wherever the highest degree of safety needs to be guaranteed for humans, machines and the environment. A system being fail safe means, not that failure is impossible or improbable, but rather that the system design prevents or mitigates unsafe consequences of the system failure. That is, if and when a fail-safe system fails in any case, accidents and damage as a result of fault must be evaded at all costs. Thus when controlling, dangerous or critical machinery, it is necessary to device and implement a fail-safe strategy to ensure that the machine operates safely even when, the elements of the control hardware or software fail.

The types of failure are many, hence a thorough analysis of the failure mode and effects are used for identifying the failure situations and design safety procedures. Some systems can never be made fail safe, as continuous availability is needed. Redundancy, fault tolerance, or recovery procedures are employed in these situations. This also makes the system, less sensitive for the reliability prediction errors or quality induced uncertainty for the separate items. On the other hand, failure detection, correction and avoidance of CCF become increasingly important to ensure system level reliability.

1. In industrial automation, usually alarm circuits are normally closed. This ensures that in case of a wire break the alarm will be triggered. If the circuit were normally open, a wire failure would go undetected, while blocking actual alarm signals.

2. In control systems, critically important signals can be carried by a complementary pair of wires. Only states where the two signals are opposite (one is high, the other low) are valid. If both are high or both are low, the control system knows that something is wrong with the sensor or connecting wiring. Simple failure modes such as dead sensor cut or unplugged wires, are thereby detected. An example would be a control system reading both the normally open (NO) and normally closed (NC) poles of a SPDT selector switch against common, and checking them for coherency before reacting to the input.

3.4.2   FAULT TOLERANT SYSTEM

Certain utility or plant operations, once started have to continue as uninterrupted and nonstop operations, even in the case of a hardware or equipment failure. In other way, these are fault tolerant systems which has the ability of preventing a catastrophic let-down, that could result from a single point of failure. A fault-tolerant system is designed from the ground up for reliability by building multiples of all critical components, such as CPUs, memories, disks and power supplies into the same computer. In the event one component fails, another should take over without skipping a single point of operation. A feasible strategy for a fault tolerant system is by anticipating exceptional conditions and design a system to cope with them. The basic aim is that the system should be able to self-stabilize after the fault and converge towards an error free state. The above strategy may not be successful in some cases. In such applications, fault tolerance is implemented by providing redundancy.

3.4.3   GRACEFUL DEGRADATION SYSTEMS

Fault tolerance is often used synonymously with graceful degradation, although the latter is more aligned with the more holistic discipline of fault management, which aims to detect, isolate and resolve problems pre-emptively. A fault-tolerant system swaps in backup componentry to maintain high levels of system availability and performance. But Graceful degradation allows a system to continue operations, with a reduced state of performance.

3.4.4   DESIGN CONSIDERATIONS FOR FAULT TOLERANT SYSTEM

While designing a fault tolerant system, one may consider the business continuity requirements, disaster recovery plan, the disaster recovery products presently available in the market, and of course the budget and available manpower for engaging. Nevertheless Fault-tolerant systems are designed to compensate for multiple failures. The failure point has to be specifically identified, and a backup component or an immediate procedure should be taken its place with no loss of service. The failure point can be computer processor unit, I/O subsystem, memory cards, motherboard, power supply, network components in the cloud, communication servers of the service provider, etc. Hence each and every stage should be thoroughly analyzed and necessary provisions may be envisaged and implemented.

In a software implementation, the operating system (OS) provides an interface that allows a programmer to checkpoint critical data at predetermined points within a transaction. In a hardware implementation, the programmer need not to be aware of the fault-tolerant capabilities of the machine.

At a hardware level, fault tolerance is achieved by duplexing each hardware component. Disks are mirrored. Multiple processors are lock-stepped together and their outputs are compared for correctness. When an anomaly occurs, the faulty component is determined automatically, and is taken out of service, but the machine continues to function as usual.

3.4.5   HIGH AVAILABILITY

Fault tolerance is closely associated with maintaining business continuity via highly available computer systems and networks. Fault-tolerant environments are defined as those that restore service instantaneously following a service outage, whereas a high-availability environment strives for setting up of independent servers coupled loosely together to guarantee system-wide sharing of critical data and resources. The loosely coupled clusters monitor each other's health and provide fault recovery, to ensure applications remain available. Conversely, a fault-tolerant cluster consists of multiple physical systems that share a single copy of a computer's OS. Software commands issued by one system are also executed on the other system. Systems with integrated fault tolerance incur a higher cost due to the inclusion of additional hardware.

3.4.6   CRITICAL FUNCTIONS

If any operation of plant or process is critical, i.e. if the downtime costs are high or cause any human fatality or any expensive hardware destruction, redundancy must be incorporated into the system to eliminate system failures due to equipment failure. Such functionalities are categorized as mission critical functions. In fact it is the privilege of the critical function to have software and hardware redundancy to be fault tolerant in a SCADA system. In power system SCADA with remote access RTU sites, both system redundancy and channel redundancy must be ensured for critical functions. In such cases RTU with CPU redundancy and communication port redundancy are highly recommended. At the master station end, the routers, firewalls, and servers must be configured to have redundancy for a fault tolerant system. Mission-critical installations often have separate power sources in case of a power failure, and installations in areas prone to natural disasters or the threat of fire, keep the servers in different geographic locations. However, whatever type of disaster recovery is planned for, it is possible to greatly reduce lost data and downtime by planning the proper system design, and by choosing a SCADA system with built-in redundancy. While ensuring system redundancy requirement for critical functions to be fault tolerant, it is highly recommended to consider the following points.

1. Dual networks for full LAN redundancy,
2. Redundancy can be applied to specific hardware,
3. Supports primary and secondary equipment configurations,
4. Intelligent redundancy allows secondary equipment to contribute to processing load,
5. Automatic changeover and recovery,
6. Mirrored disk I/O devices,
7. Mirrored alarm servers, and
8. File server redundancy.

A typical redundant and fail safe DCS master control station configuration with HA is shown in Figure 3.7 below. The sub components such as DMZ, SCADA Control System LAN, Dispatch Training Simulator, and Fail Safe connectivity to Smart Grid MCC using HA are shown separately in Figure 3.8, Figure 3.9, Figure 3.10 respectively. Here a De Militarized Zone (DMZ) is designed carefully and kept exclusively between two Firewalls of different make. The session from the servers of DMZ to the servers of SCADA zone is strictly made forbidden by appropriately defining the Firewall ruleset.

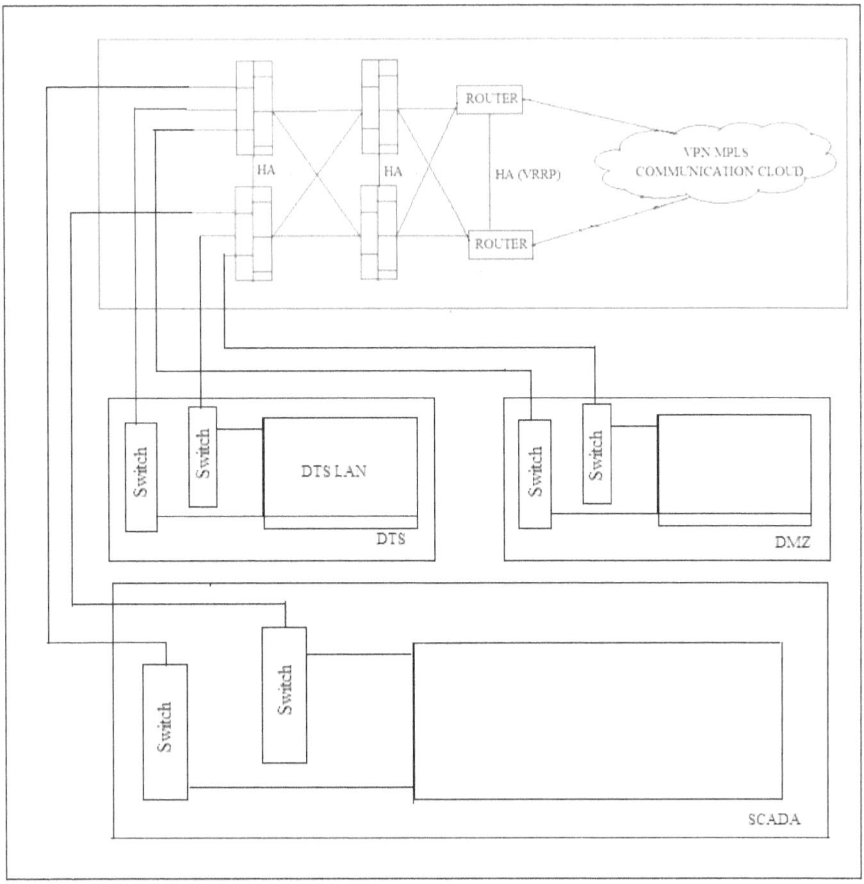

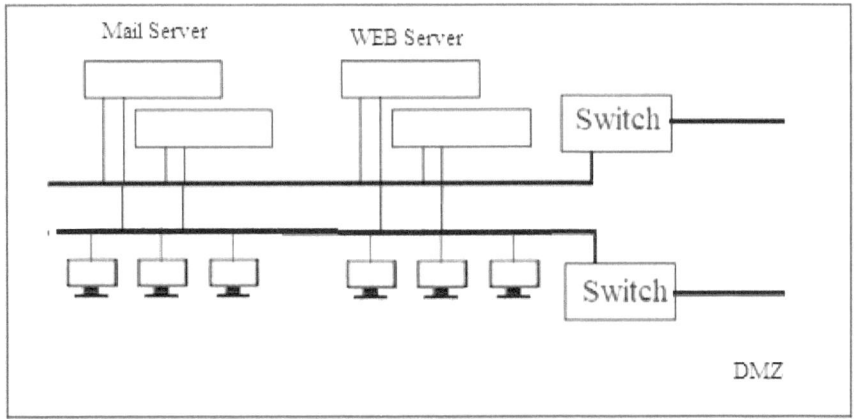

Figure 3.7   Block diagram of a typical DCS MCC with Fail Safe HA connectivity

Figure 3.8   De Militarized Zone (DMZ)

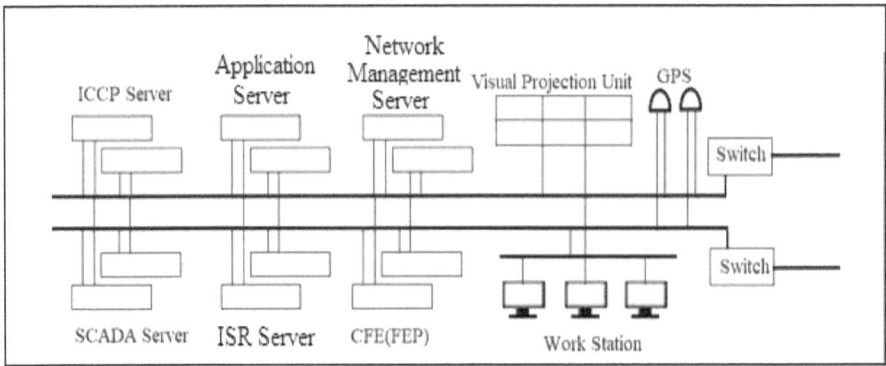

Figure 3.9   SCADA Control System LAN

Dispatcher Training Simulator (DTS) is a subsystem within Distribution Management System (DMS) that operates separately from the real–time system and provides a realistic environment for hands–on dispatcher training under simulated normal, emergency, and restorative operating conditions. The training is based on interactive communication between instructor and trainee. The DMS training simulator serves two main purposes:

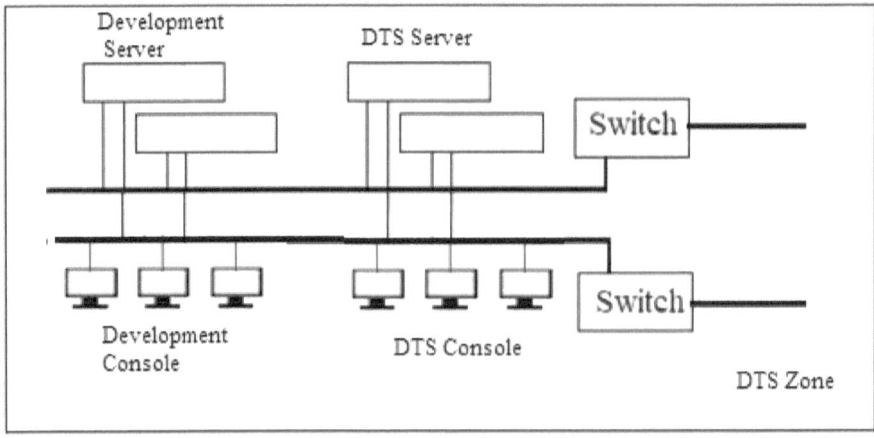

Figure 3.10   Dispatch Training Simulator

1. Allowing personnel to become familiar with the DMS system and its user interface without impacting actual substation and feeder operations.

2. Allowing personnel to become familiar with the dynamic behavior of the electric distribution system in response to manual and automatic actions by control and protection systems during normal and emergency conditions.

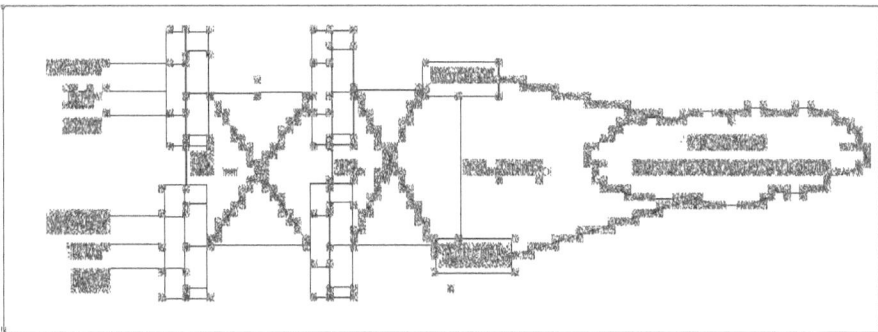

Figure 3.11   Fail Safe connectivity to DCS MCC using HA

The Figure 3.11 descibes a Fail Safe connectivity to DCS MCC using HA which is mostly preferred in the Power System SCADA and DCS which control the critical operations.

3.4.7   SYSTEM REDUNDANCY

Redundancy is the hallmark of fault tolerant systems. It is defined as additional or alternative systems, sub-systems, assets, or processes that maintain a degree of overall functionality in case of loss or failure of another system, sub-system, asset, or process. Normally power system SCADA redundancy, starts from the data capturing units such as RTU or PLC. If the SCADA is a distributed system, then every system component upto the MODEM have to be redundant. Further if the communication to the master control station is by means of unguided media, then system should have antenna redundancy as well. Obviously this adds cost considerably and the implementing agencies/vendors normally find pretexts for compromise which is the general lapse observed in implementing power system SCADA. But this makes the SCADA system unreliable.

Definitely true fault tolerant SCADA systems with redundant hardware are the most costly as the additional components add to the overall system cost. However, fault tolerant systems provide the same processing capacity after a failure as before, and ensures safety.

3.4.8   CHANNEL REDUNDANCY

Most of the power system SCADA is distributed in nature and geographically separated. Hence remote site to site communications are required. This mainly depends on third party communication service providers.

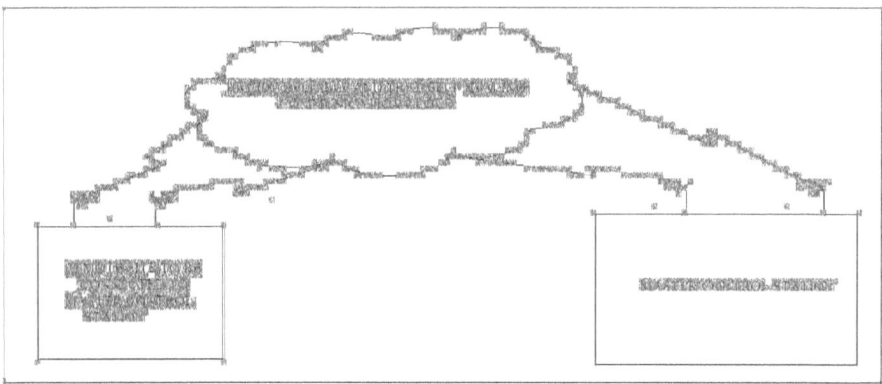

Figure 3.12a   Channel redundancy achieved with a resilient and reliable cloud

While selecting the third party communication providers, channel redundancy has to be ensured to have true system redundancy. Mainly this is achieved in two ways which is explained below

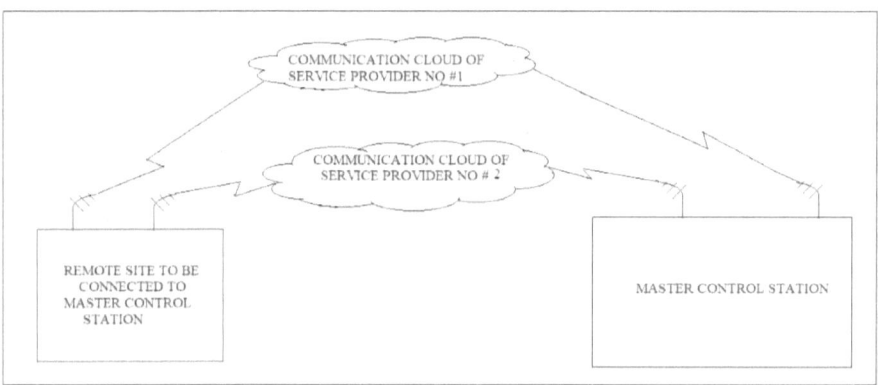

Figure 3.12b   Channel redundancy achieved with two different communication clouds

If the third party communication provider is highly reliable and having a true self healing communication cloud as per the requirement of the utility, then connectivity to the two geographically different nodes is an acceptable solution on economic reasons. However in the strictest sense, true channel redundancy for critical functions ideally demands two different communication providers with active-active or active-standby mode. These are shown in Figure 3.12a and 3.12b respectively.

3.5 DESIGN AND CONFIGURATION CONSIDERATIONS OF MCC

The following aspects are utmost important while designing the Master Control Center of the power system SCADA.

1. Sufficient hardware and software redundancy have to be ensured, to achieve the overall system redundancy with High Availability having No Single Point of Failure (NSPF) as the SCADA functions for the power system are critical. This includes communication channel as well,

2. The firewall must be properly configured with the appropriate ruleset, after due deliberation with the security policy of the utility. Design with a two layer Firewall configuration, with different make IDS loaded are not a better option but a must, while moving for a DMZ, and

3. The CFE or the FEP of the SCADA control center must have high end capabilities with suitable IDS loaded and have cryptographic capabilities. If third party communication media are used, especially using the VPN for data transfer, utmost care must be given for proper VPN termination.

SUMMARY

This chapter begins with describing the communication architecture of basic SCADA. Then moves on to describing the common communication philosophies adopted in DCS. As the reliability and availability of DCS functions are the most important, they are briefly introduced, but cater the necessary understanding to power system professionals and engineering students who are engaged or intend to embark into the DCS, Smart Grid or Microgrid domain. It then explains the concepts of Fault Tolerant Systems, Fail Safe and redundant systems, High Availability, etc. Based on these concepts, this chapter then elaborates the design of a typical SCADA MCC architecture having redundant and HA connectivity.

Chapter FOUR

SCADA APPLICATIONS

4.1   INTRODUCTION

Today the word SCADA became almost synonym to automation as it even permeates the nation's critical infrastructures in addition of monitoring and controlling a variety of plants, processes and operations in different areas from chemical, gas, water, communications, automobiles and power systems. The basic SCADA architecture is very similar, though the application area varies. The rapid developments in the communication technology broke the geographical barriers and elevated SCADA to another level of remote monitoring and control. Obviously it invited cyber security threats which necessitate a secure SCADA (sSCADA) technology. This chapter elaborates a few areas where SCADA applications are invariably deployed to improve efficiency in many aspects.

4.2   POWER SECTOR

Power utilities uses the SCADA systems to detect current flow and line voltage, to monitor the operation of circuit breakers, and to take sections of the power grid offline and online. The Energy Management System (EMS) and Distribution Management System are the two major SCADA applications. Further SCADA is the heart and brain of the Smart Grid and microgrid technologies. A fairly elaborated description of the EMS and DMS is given in the succeeding sections.

4.2.1 ENERGY MANAGEMENT SYSTEMS (EMS)

With deregulation of the power industry and with the development of the Smart Grid, decision-making is becoming decentralized, and coordination between different actors such as Independent System Operators (ISO), Regional Transmission Operators (RTO), Distribution Generators (DG), including Renewable Energy Sources (RES), energy traders, and prosumers, in various markets becomes important. All these complex changes, and the aging infrastructure which limits the operating range and increase in congestion of the networks, needs a modern management and control system to carry out all these functions with optimum efficiency. This leads to the development of Energy Management System (EMS) which make use of SCADA technology. A block diagram of EMS is shown in Figure 4.1.

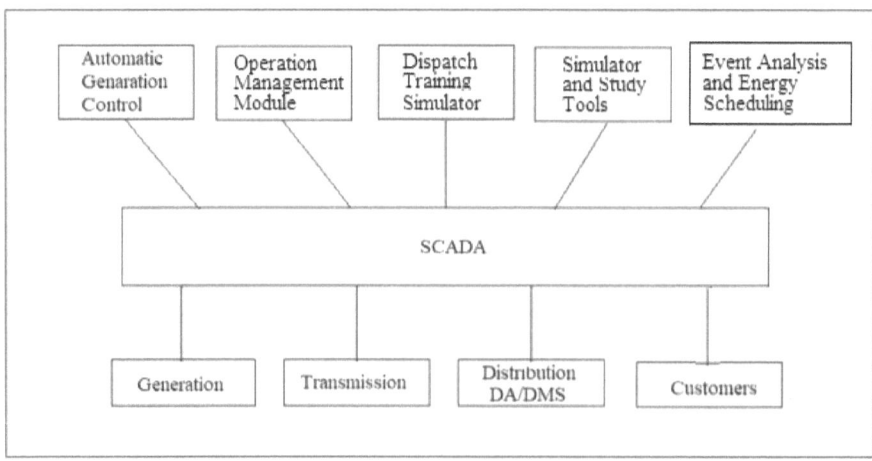

Figure 4.1   Block Diagram of an EMS

EMS is a suitable blend of software tools and high end hardware, preferably, high end machines configured with High Availability (HA) for reliable and efficient operation of generation and transmission assets of a power utility in real-time. In fact EMS monitor, control and optimize the power transmission and generation in a stringent secure manner.

Hardware part of EMS consists of RTU, Intelligent Electronic Device (IED), Control and Protection, Computer networking, etc. System status and measurement information are collected by the RTUs and sent to the

Control Centre through the communication infrastructure. The CFE in the EMS is responsible for communicating with the RTUs and IEDs. Different EMS applications reside in different servers and are linked together by the LAN. Software part of EMS consists of Application programs for network analysis of power systems. In EMS, application programs are run in a real-time as well as extended real-time environment to keep the power system in a secure operating state. EMS application has been briefly described below.

Real-time SCADA Applications: Provides Supervisory Control and Data Acquisition including alarm or events, tagging, data historians, data links, control sequences, and load shed applications used to monitor or operate the network.

Generation Dispatch and Control: GDC provides the functions required for dispatch and closed loop digital control of multiple generators in an economic fashion while adhering to standard operating guides at the same time considering interchange schedules, dynamic schedules (load or generation in an out of the area), inadvertent interchange payback, time error correction, reserve requirements, and security constraints of the transmission network.

Transmission Security Management: TSM provides sophisticated applications to analyze and optimize the use of the transmission network in a reliable and secure manner.

Energy Scheduling and Accounting: ESA provides applications to monitor standard reporting criteria, production costs, interchange scheduling, inadvertent interchange accounting, and weather adaptive demand forecasting.

Control functions: It includes real-time monitoring and control functions such as Automatic Control and automation of a power system, Efficient Automatic Generation Control (EAGC) and Load Frequency Control (LFC), Reactive Power Control (RPC), Reactive Power and Voltage Control and Optimal automatic generation control across multiple areas, and Tie -line control.

Operating functions: Ensure Economic and optimal Operation of the generating system, Supports Efficient operator Decision Making, Improved power quality of supply and optimization functions such as

1. Assists the operator for optimal utilization of the transmission network,
2. Assists Power scheduling interchange between areas,
3. Ensure Optimal allocation of resources, and
4. Real-time overview of the power generation interchanges, and reserves.

Planning functions: These are improved power quality of supply and system reliability, forecasting of loads and load patterns, generation scheduling based on load forecast and trading schedules, maintaining reserves and committed transactions, and calculation of fuel consumption, production costs and emissions.

Protection and Security Functions: Protective relaying, primary protection and secondary protection or backup protection are the protection functions of EMS. The protective functions are the first which are activated in a real-time operation as a protective measure. These are followed by essential control functions like LFC, AGC, RPC, etc..

When the system is insecure, security analysis informs the operator which contingency, is causing insecurity and the nature and severity of the anticipated emergency. Besides these EMS functions, a training tool, the Dispatcher Training Simulator, is embedded within an EMS. Dispatch Training Simulators were originally created as a generic application to introduce operators to the electrical and dynamic behavior of a power system. Today, they model the actual power system being controlled with reasonable fidelity and are integrated within the EMS to provide a realistic environment for operators and dispatchers to practice normal, everyday operating tasks and procedures as well as experiencing emergency operating situations. Various training activities can be practiced safely and conveniently with the simulator responding in a manner similar to the actual power system.

4.2.2   Distribution Management System (DMS)

Distribution Management System (DMS) is a set of application software and systems. Further it helps to manage distribution assets and plays an important role in ensuring power quality in power distribution, making use of real-time data acquired from distribution SCADA, Customer

Information Systems (CIS), and Geographical Information Systems (GIS). Present day DMS operates jointly with Outage Management System (OMS) and Asset Management System (AMS) for optimum reliability and efficiency. Proper distribution and sub-station sensing, and automation can reduce outage and repair time, maintain voltage level and improve asset management. Advanced distribution automation processes real-time information from sensors and meters for fault location, automatic reconfiguration of feeders, voltage and reactive power optimization, or to control distributed generation. Sensor technologies can enable condition and performance based maintenance of network components, optimizing equipment performance and hence effective utilization of assets. In the smart grid scenario, AMI is an integral part of the DMS for bi-directional information sharing between the utility and the consumer. The following are some of the advantages of implementing the SCADA/DMS in Distribution Automation.

1. Quick isolation of faulty section and fast restoration of healthy section so that, only least customers are affected during outage period.

2. All data are available in real-time. Historical data can be archived for planning. These data can be shared with all stakeholders and MIS

3. Implementation of SCADA/DMS requires one time capital investment. But a robust SCADA/DMS system, implemented in a phased manner can bring returns within a short period.

4.2.2.1 Distribution Network (DN) Model or Dynamic Mimic Diagram

In a Distribution Automation, the various components distribution system include all the primary substation feeders, distribution network and devices. The devices which generally represented in the Dynamic Mimic Diagram (DMD) are Power Injection points, Transformers, Feeders, Balanced and Unbalanced Load, Circuit Breakers, Sectionalizes, Isolators, Fuses, Capacitor banks, Reactors, Generators, Bus bars, Temporary Jumper, Cut and Ground, Meshed and Radial network configuration, Line segments, which can be single-phase, two-phase, or three phase, Conductors, Grounding devices, Fault detectors, IEDs, Operational limits for components such as lines, transformers, and switching devices, etc. The database of the network model of the utility system having an interface with the Graphical User Interface (GIS) system of the area which can give a rich visual network presentation for crew management and asset information.

4.2.2.2   Network Connectivity Analysis (NCA) or Topology Analyzer (TA)

The network connectivity analysis function provides the connectivity between various network elements. The prevailing network topology will be determined from the status of all the switching devices such as circuit breaker, isolators etc. that affect the topology of the network modeled. NCA runs in real-time as well as in study mode. Real-time mode of operation uses data acquired by SCADA. Study mode of operation will use either a snapshot of the real-time data or save cases. NCA can run in real-time on event-driven basis. The network topology of the distribution system will be based on

1. Tele-metered switching device statuses,
2. Manually entered switching device statuses, and
3. Modeled element statuses from DA applications.

The NCA assists the power system operator to know the operating state of the distribution network giving details of loops and parallels in the network.

4.2.2.3   State Estimation (SE)

Network state comprises the set of voltage phasors for all buses. State variables comprise all other variables such as voltages drops, section currents, and load which can be calculated from the Network state. SE represents the basic DMS application, since practically all other DMS applications are based on its results.

The State Estimation (SE) which is used for assessing the distribution network state uses the data acquired remotely from field devices. The state estimator converts the acquired data and the generated measurement data into consistent set of network states and state variables viz voltage phasor of the network data and network topology, historical and static consumer data. In fact state estimator assess loads of all network nodes, and, subsequently, assess all other state variables such as voltage and current phasors of all buses, sections and transformers, active and reactive power losses in all sections and transformers in the Distribution network.

4.2.2.4   Volt -VAR Control (VVC)

In electrical power system the reactive power can be generated at source generators or can be injected at the substations through Volt-VAR systems. It is more appropriate to inject at substations rather than producing then at generator points and transporting them over long distances. Any power system always tries to optimize on the reactive power flow over their networks. The coordination of voltages and reactive power flow control requires, coordination of VOLT and the VAR function. This function shall provide high-quality voltage profiles, minimal losses, controlling reactive power flows, minimal reactive power demands from the supply network. TAP Changer for voltage control and VAR control devices such as switchable and fixed type capacitor banks are the main resources to be considered in any voltage and reactive power flow control.

4.2.2.5   Load Flow Application (LFA)

Load flow studies, identify line loads and bus voltages which are out of range and inappropriately large bus phase angles and other parameters having the potential to create operating difficulties. Intermediate load and off-peak load studies are also useful, as off-peak loads can result in high voltage conditions that are not identified during peak loads. Load flow studies assist system operators in calculating power levels at each generating unit for economic dispatch, analyzing outages and other forced operating conditions, and coordinating power pools. In most instances, load flow studies are used to assess system performance and operations under a given condition. The Load Flow function provides

1. real or active and reactive losses on the Station power transformers, feeders and distribution circuits including feeder regulators and distribution transformers,

2. phase and neutral currents for each feeder, three phases and per phase KW and KVAR losses in each feeder, section, Distribution Transformer,

3. active and reactive power flows in all sections, overloaded feeders, lines, busbars, transformers loads etc including the actual current magnitudes,

4. the overload limits with feeder and substation name, and

5. limit violations of voltage magnitudes, overloading, voltage drops, etc.

4.2.2.6   Load Shed Application (LSA)

Load Shed Application (LSA) is aimed for shedding of load under emergency conditions, as well as restoring load after restoring system conditions when demand is higher than supply due to several reasons such as the faults, tripping of lines, insufficient generation, etc.

In these situations the power system operator tries to distribute available power through shedding of loads to consumers over small definite periods till he tides over the situation of loss of power. The load-shed application helps to implement an autonomous load shedding system to automate and optimize the process of selecting the best combination of switches to be opened and controlling in order to shed the desired amount of load. Given a total amount of load to be shed, the load shed application shall recommend different possible combinations of switches to be opened, in order to meet the requirement. The SCADA DMS operator presents with various combinations of switching operations, which shall result in a total amount of load shed. The despatcher can then choose the most suitable actions recommended and execute it. In reality, in case of failure of supervisory control for few breakers, the total desired load shed may not be met. Under such conditions, the application will inform the operator regarding the balance amount of load to be shed. The load-shed application runs again to complete the desired load shed process.

4.2.2.7   Fault Management and System Restoration (FMSR) Application

The availability of data related to the breaker status and the level of the fault current flowing in the networks helps to manage and restore the system in an event of fault. Thus SCADA DMS application helps to provide the assistance to the power system despatcher to detect, localize, isolate and restore the distribution system after a fault in the system has occurred. The devices which help in localisation and isolation of the fault includes, Auto Reclosures (AR), Sectionalisers, Fault Passage Indicators etc.

4.2.2.8   Loss Minimization via Feeder Reconfiguration (LMFR)

This is one of the major applications of the SCADA DMS. LMFR is also a switching operation during fault, which requires to supply power through alternate feeders in the distribution network, by modifying the feeder configuration topology. The information of network topology and availability of adjacent feeder networks are well utilized to select an alternate optimum feeder path with overall aim of reducing the line losses and maximum quality power delivery to the consumers. This function also explores the opportunities to minimize technical losses in the distribution system with a proper reconfiguration of feeders in the network for a given load scenario. In addition it helps to calculate the current losses based on the loading of all elements of the network. The telemetered data, which are not updated due to telemetry failure, can also be considered by LMFR application based on arriving at the recommendations of LFA. The LMFR application can be utilized to have the various scenarios for a given planned and unplanned outages.

4.2.2 9   Load Balancing via Feeder Reconfiguration (LBFR)

The discussions had on previous topic can be used for the Load Balancing via Feeder Reconfiguration for the optimal balance of the segments of the network that are over and under loaded. This helps in better utilization of the capacities of distribution facilities such as transformer and feeder ratings. The Feeder Reconfiguration Function can also be used to have a scenario on an overload condition, unequal loadings of the parallel feeders and transformers, periodically or on demand in the network by the despatcher. The system will help generating the switching sequence to reconfigure the distribution network for transferring load from some sections to other sections. The LBFR application can even consider the planned and unplanned outages, equipment operating limits, tags placed in the SCADA system while, recommending the switching operations. The function helps in distributing the total load of the system among the available transformers and the feeders in proportion to their operating capacities, considering the discreteness of the loads, available switching options between the feeder and permissible intermediate overloads, during switching. The despatcher can have the options to simulate switching operations and visualise the effect on the distribution network by comparisons based on line loadings, voltage profiles, load restored, system losses, number of affected customer,etc.

4.2.2.10   Distribution Load Forecast (DLF)

The Distribution Automation system keeps logging of network data periodically. This historical database and weather conditions data collected over a period can be used for prediction and to have forecasting of the requirement of consumer loads. Generally there are two types of forecasting viz. Short Term Load Forecasting (STLF) and Long Term Load Forecasting (LTLF). STLF will be used for assessment of the sequence of average electrical loads in equal time intervals, from 1 to 7 days ahead. The LTLF is used for forecasting load growths over longer durations. The forecasting techniques are based on different forecasting methods such as

1. Autoregressive,
2. Least Squares Method,
3. Time Series Method,
4. Neural Networks,
5. Kalman filter, and
6. Weighted Combination of these method.

4.2.2.11   Outage Management System (OMS)

An outage management system is a network management software that is capable of restoring the network model after an outage. Outage management systems are integrated tightly, resulting in timely and accurate actions along with supervisory control. Outage management systems are not only capable of performing restoration activities related to service, but also capable of tracking, displaying and grouping outages.

An OMS is mainly used by operators in electric distribution systems. It can help in identifying the portion of the circuit responsible for the interruption. Based on the different criteria present in the network, it can also assist in grouping and prioritizing the resources and indirectly help in minimizing the impacts of outages.

An outage management system has the following features:

1. Prioritizes restoration efforts and management of resources upon outages,
2. Provides supervisors with an estimated timeline of restoration,
3. Reports the actual cause of the outage, and
4. Provides accurate information about the extent of the outage and its impact on customers and their management

The benefits of outage management software are mentioned below.

1. Prioritization of resources and planning involved in outage management software results in reduced outages and faster recovery.
2. Customer relationship is improved due to better management of outage issues.
3. Because of the tracking involved, there is better prediction of outages, allowing them to be handled properly.
4. Operational efficiency is increased compared to situations where an outage management system is not in place.
5. Operational visibility across the network increases greatly with the use of an outage management system.
6. Decision making is faster for supervisors because of the reports provided by the application, even in cases of complex outages.

4.3   OIL AND GAS INDUSTRY

One of the major areas where SCADA system has been employed is Oil and Gas Industry, especially for petroleum refining. A pictorial representation of the refining process is given in Figure 4.2.

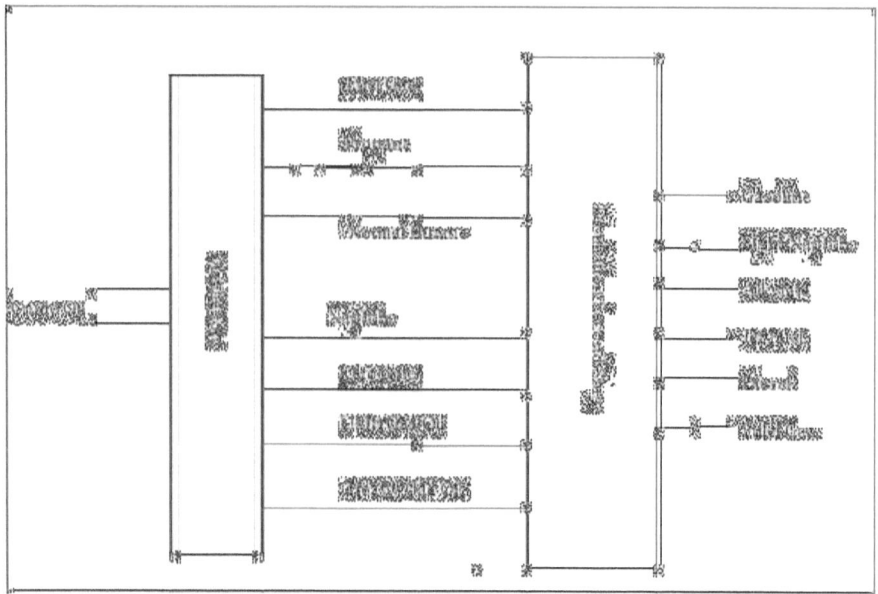

Figure 4.2   Petroleum Refining Process

Many complex parameters such as temperature, pressure, flow, density, etc. are to be monitored in different levels and to be controlled to obtain the petroleum products in correct proportion. Mostly petroleum products requires a local SCADA where the field devices are connected through a field bus. The entire refining process is controlled by SCADA, by monitoring various parameters centrally to yield the various products and its biproducts.

4.4   AUTOMOBILE INDUSTRY

At present the automobile industry is undergoing tremendous changes with the integration of modern automation technology. Almost every function of the automobile is being automated. The driver less vehicles are no longer a dream, but a reality. In all these technological innovations, SCADA plays the pivotal role of automation. The data inputs required for the control and command are captured through the dedicated transducers, communicated to the Engine Control Unit (ECU) in a secure and safe manner. So today SCADA in addition to automate the automobile industrial process, it find applications within the automobile itself.

4.5   WATER DISTRIBUTION SECTOR

4.5.1   WATER PUMPING STATIONS

Usually water pumping is carried out as per the availability of quantity of water at different sources. If sufficient water is not available in the wells then it pump from ground water source. The master pumping station, monitor the availability of water in all the water sources through the SCADA field devices and verify the availability to meet the present requirement. Then it prepares a scheme of pumping from various sources to meet the requirements with the limits and constrains. Often it may be presented or communicated to the operator concerned and necessary approval may be obtained before it starts pumping.

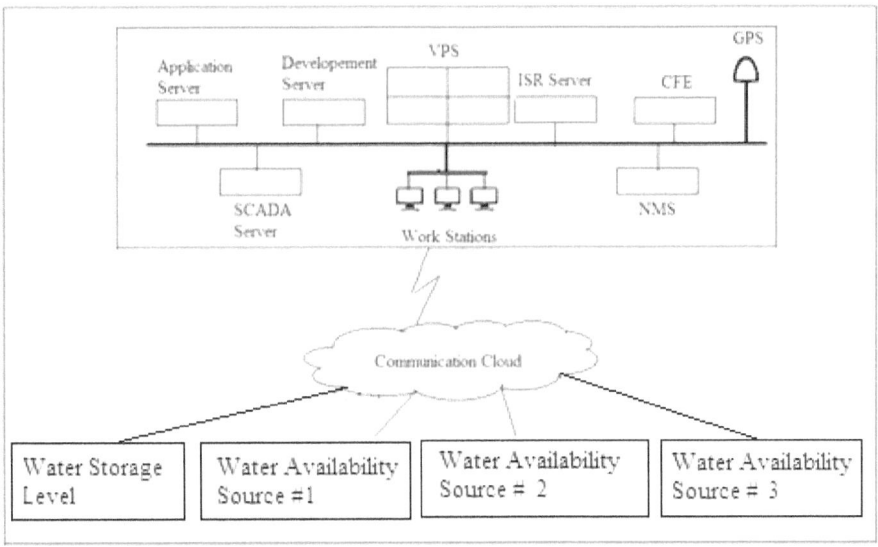

Figure 4.3   SCADA for Water Pumping

In case the water quantity cannot be met from the normal surface water sources, then appropriate decision to pump it from ground water sources will be taken. Once the required quantity water is pumped and stored in the reservoir, necessary commands are issued to the various pumping stations to turn off the pumping. In fact the SCADA system coordinates various pumping resources and implement an automated efficient and economical pumping and storage system.

4.5.2   DISTRIBUTION PIPELINE PRESSURE MONITORING AND CONTROL

In a water distribution system, the major objective is to ensure 24×7 quality water availability to the consumers. But it may not be always possible due to many reasons such as non availability of water sources to meet the required quantity, due to fault in the distribution network and pumping stations. Further in certain situations, specific areas like hospitals has to be ensured with 24×7 quality water availability, despite water scarcity. In these circumstances, an automation using the SCADA technology is an ideal solution.

In automated water distribution system, pressure sensors are suitably mounted in various locations, to monitor the pressure at strategic points. The valves with remote controls which deployed appropriately can be controlled from the MCC through SCADA to meet the required distribution strategy as shown in Figure 4.4.

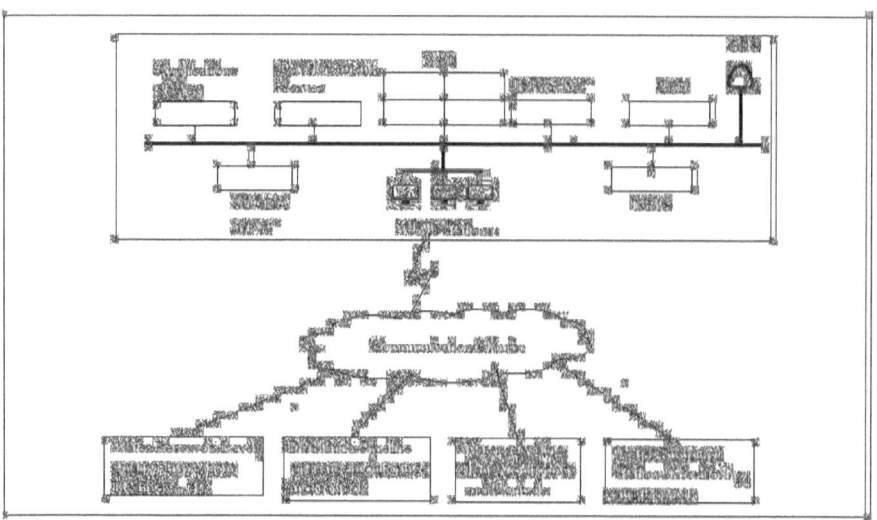

Figure 4.4   SCADA for Water Distribution

4.5.3   WATER RECYCLING PLANT MONITORING AND CONTROL

In addition to these applications, SCADA is being extensively used in building facilities and environments by the facility managers to control HVAC, refrigeration units, lighting and entry systems. In manufacturing,

SCADA systems manage parts inventories for just-in-time manufacturing, regulate industrial automation and robots, and monitor process and quality control. In mass transit and railway traction, transit authorities use SCADA to regulate electricity to subways, trams and trolley buses, to automate traffic signals for rail system, to track and locate trains and buses, and to control railroad crossing gates.

SUMMARY

This chapter begins with a detailed description of the SCADA applications in power sector viz the Energy Management System and Distribution Management System. Then it gives a brief description of SCADA applications in water pumping, water distribution, and water treatment. Attempts has also been made to briefly explain the SCADA applications in public transportation, automobile, and oil and gas industry.

Chapter FIVE

ADVANCED SCADA COMMUNICATIONS

5.1 INTRODUCTION

The key enabler for the ICS and DCS is the availability of secure and preferably bi-directional data communications and the proper amalgamation of distributed intelligence and communication technologies. Thus, the design and implementation of a modern, reliable communications infrastructure is a fundamental and important requirement for making the ICS or DCS smarter.

The medium by which data is transmitted is known as a communication channel or communication media. The transfer of data takes place in the form of analog signals and then it is measured in the form of bandwidth, the higher the bandwidth the more the data that will be transferred. Communication media are broadly classified into two categories, namely, guided media (wired) and unguided media (wireless). Both are used for short distance (LANs, MANs) and long distance (WANs) communication. This chapter elaborates various guided and unguided media used in communication, its merits and demerits, and practical considerations to be taken care while selecting the media for different applications. The chapter also explains the various communication technologies currently available for the deployment in ICS and DCS. Before that, it is better to be familiar with certain terminologies which are briefly explained below.

5.2 TYPES OF TRANSMISSION

There are various transmission classifications depending upon the different technologies employed and they are mainly based on

1. Analog and digital,

2. Synchronous and Asynchronous,

3. Broadcast, Multicast, and Unicast,

4. Simplex, Half Duplex, and Full Duplex, and

5. Baseband and Broadband.

As the communication is the key enabler of the modern SCADA, a basic understanding of these terms become most essential for an automation engineer. Hence brief explanation of these terminologies is given below.

5.2.1 ANALOG AND DIGITAL

There are two types communication technologies in practice and they are analog and digital. Analog communications occur with a continuous signal that varies in frequency, amplitude, phase, voltage, and so on. The variances in the continuous signal produce a wave shape as opposed to the square shape of a digital signal. Digital communications occur through the use of a discontinuous electrical signal and a state change or ON-OFF pulses. In other way the information is encoded digitally as discrete signals and transmitted electronically to the recipients. Digital signals are more reliable than analog signals over long distances or when interference is present. This is because of a digital signal's definitive information storage method employing direct current voltage where voltage ON represents a value of 1 and voltage OFF represents a value of 0. These ON-OFF pulses create a stream of binary data. Analog signals become altered and corrupted because of attenuation over long distances and interference. Since an analog signal can have an infinite number of variations used for signal encoding as opposed to digital's two states, unwanted alterations to the signal make extraction of the data more difficult as the degradation increases. A brief comparison between these two technologies is summarized below.

1. *Bandwidth:* This factor creates the key difference between Analog and digital communication. Analog signal requires less bandwidth for the transmission while digital signal requires more bandwidth for the transmission,

2. *Power Requirement:* Power requirement in case of digital communication is less when compared to analog communication. Since the bandwidth requirement in digital systems is more, they consume less power,

3. *Fidelity:* Fidelity is a factor which creates a crucial difference between Analog and digital communication. Fidelity is the ability of the receiver which receives the output exactly in coherence with that of transmitted input. Digital communication offers more fidelity as compared to Analog Communication,

4. *Noise Distortion and Error Rate:* Analog systems are affected by Noise while, digital systems are immune from Noise and Distortion. Error rate is another significant difference which separates Analog and Digital Communication. In Analog instruments, there is an error due to parallax or other kinds of observational method,

5. *Synchronization:* Digital communication system offers to synchronize which is not effective in analog communication. Thus, synchronization also creates a key difference between Analog and Digital Communication,

6. *Cost:* Digital communication equipments are costly and digital signal require more bandwidth for transmission, and

7. *Hardware Flexibility and Portability:* The hardware of analog communication system is not as flexible as digital communication. Analog systems are less portable, as components are heavy while digital systems are more portable as they are compact equipments.

5.2.2   SYNCHRONOUS AND ASYNCHRONOUS

Communications are either synchronous or asynchronous. Some communications are synchronized with some sort of clock or timing activity and referred to synchronous communications. It relies on timing or clocking mechanism based on either an independent clock or a time stamp embedded in the data stream. Synchronous communications are typically able to support very high rates of data transfer. On the other hand, asynchronous communications rely on a stop and start delimiter bit to manage the transmission of data. Because of the use of delimiter bits and the stop and start nature of its transmission, asynchronous communication is best suited for smaller amounts of data. Public Switched Telephone Network (PSTN) modems are good examples of asynchronous communication devices.

5.2.3   BROADCAST, MULTICAST, AND UNICAST

Broadcast, multicast, and unicast technologies determines the number of destinations a single transmission can reach.

Unicast technology supports only a single communication to a specific recipient. In this case there is just one sender, and one receiver. Unicast transmission, in which a packet is sent from a single source to a specified destination, is still the predominant form of transmission on LANs and within the Internet. All LANs and IP networks support the unicast transfer mode, and most users are familiar with the standard unicast applications (e.g. http, SMTP, FTP and telnet) which employ the TCP transport protocol.

Broadcast technology supports communications to all possible recipients. In this case there is just one sender, but the information is sent to all connected receivers. Broadcast transmission is supported on most LANs (e.g. Ethernet), and may be used to send the same message to all computers on the LAN (e.g. the Address Resolution Protocol (ARP)) uses this to send an address resolution query to all computers on a LAN. Network layer protocols (such as IPv4) also support a form of broadcast that allows the same packet to be sent to every system in a logical network (in IPv4 this consists of the IP network ID and an all 1's host number). One example of an application which may use multicast is a video server sending out networked TV channels. Simultaneous delivery of high quality video to a large number of delivery platforms will exhaust the capability of even a high bandwidth network with a powerful video clip server. This poses a major scalability issue for applications which requires sustained high bandwidth. One way to significantly ease scaling to larger groups of clients is to employ multicast networking.

Multicast networking technology supports simultaneous communications to multiple specific recipients. In this case there may be one or more senders, and the information is distributed to a set of receivers. IP multicast provides dynamic many-to-many connectivity between a set of senders (at least 1) and a group of receivers. The format of IP multicast packets are identical to that of unicast packets and are distinguished only by the use of a special class of destination address, which, denotes a specific multicast group. Since TCP supports only the unicast mode, multicast applications must use the UDP transport protocol.

Unlike broadcast transmission, multicast clients receive a stream of packets only if they have previously elected to do so. The routers in a multicast network find the sub-networks which have active clients. Each multicast group attempt to minimize the transmission of packets across parts of the network for which there are no active clients.

The multicast mode is useful if a group of clients require a common set of data at the same time, or when the clients are able to receive and store (cache) common data until needed. Where there is a common need for the same data required by a group of clients, multicast transmission may provide significant bandwidth savings (up to 1/N of the bandwidth compared to N separate unicast clients).

The majority of installed LANs are able to support the multicast transmission mode. Shared LANs inherently support multicast, since all packets reach all the Network Interface Cards (NIC) connected to the LAN. The earliest LAN network interface cards had no specific support for multicast and introduced a big performance penalty by forcing the adaptor to receive all packets (promiscuous mode) and perform software filtering to remove all unwanted packets. Most modern network interface cards implement a set of multicast filters, relieving the host of the burden of performing excessive software filtering.

5.2.4 SIMPLEX, HALF DUPLEX, AND FULL DUPLEX COMMUNICATION CHANNELS

In a communication system, there will be a transmitter and a receiver. In between the transmitter and the receiver, there is a transmission medium of the data/information, usually referred as the communication channel. Although the required information for transmission originates from a single source, there may be more than one destination or receivers. This depends upon the number of receiving stations which are linked to the channel and the quantity of energy that the transmitted signal possesses. If the channel length is more and the transmission power is less, the receiver situated at a long distance cannot receive the data properly. In a digital communications channel, the information can be represented by a stream of bits called bytes. A collection of bytes can be grouped to form a frame or other higher-level message unit. These types of multiple levels of encapsulation facilitate the handling of messages in a complex data communications network. If any communications channel is considered, it has a direction associated with it.

Simplex Channel: It is conventional that the message source is the transmitter, and the destination is the receiver. A channel whose direction of transmission is unchanging is called as a simplex channel. In other words, a type of data transmission, where message transmission is taken place only in one direction, typical example is the radio station which is a simplex channel because it always transmits the signal to its listeners and never allows them to transmit back. Another example is the television. The advantage of simplex mode of transmission is, since the data can be transmitted only in one direction, the entire band width can be used.

Half Duplex Channel: A half-duplex channel can be considered as a single physical channel in which the direction may be reversed. Messages can flow in two directions in a half-duplex type, but never at the same time. In other words it can be said that at a single time, the transmission of data are done in only one direction. For example, in a telephone call, one party speaks while the other listens. After a pause, the other party speaks and the first party listens. Speaking simultaneously will result in a garbled sound that cannot be understood. The main difficulty of half-duplex mode of transmission is since two channels are used, the band width of the channel would be decreased.

Full Duplex Channel: A full-duplex channel can be used for bi-directional communication. In fact, message can be transmitted simultaneously in both directions. It comprises of two simplex channels, a forward channel and a backward (reverse) channel, linking at the same points. The transmission rate of the reverse channel will be very slow if it is used only for flow control of the forward channel. The main problem of the full duplex mode of transmission is, since two channels are required, the band width would be decreased.

5.2.5   BASEBAND AND BROADBAND

The number of channels that can be pushed into a single wire simultaneously over a cable segment depends on whether the customer uses baseband technology or broadband technology. Baseband technology can support only a single communication channel. Analog transmission means that data is being moved as waves, and digital transmission means that data is being moved as discrete electric pulses. Baseband uses a direct current applied to the cable. A current that is at a higher level represents the binary signal of 1, and a current that is at a lower level represents the binary signal of 0. Baseband is a form of digital signal. Ethernet is a baseband technology.

Broadband technology divides the communication channel into individual and independent subchannels so that different types of data can be transmitted simultaneously. Broadband uses frequency modulation to support numerous channels, each supporting a distinct communication session. Broadband is suitable for high throughput rates, especially when several channels are multiplexed. Broadband is a form of analog signal. As an example, a coaxial cable TV (CATV) system is a broadband technology that delivers multiple television channels over the same cable. This system can also provide home users with Internet access, but these data are transmitted at a different frequency spectrum than the TV channels. A Digital Subscriber Line (DSL) uses one single phone line and constructs a set of high-frequency channels for Internet data transmissions. A cable modem uses the available frequency spectrum that is provided by a cable TV carrier to move Internet traffic to and from a customer premise. Mobile broadband devices implement individual channels over a cellular connection, and Wi-Fi broadband technology moves data to and from an access point over a specified frequency set. Characteristics of baseband and broadband are summarized below.

Baseband:

1. Digital signals are used
2. Frequency division multiplexing is not possible
3. Baseband is bi-directional transmission
4. Short distance signal travelling
5. Entire bandwidth of the cable is consumed by a single signal in a baseband transmission.

Broadband:

1. Analog signals are used
2. Transmission of data is unidirectional

3. Signal travelling distance is long

4. Frequency division multiplexing is possible

5. Multiple frequency signals are sent simultaneously in broadband transmission.

5.3   GUIDED MEDIA

Guided media are more commonly known as wired media or bounded media, or those media in which electrical or optical signals are transmitted through cables or wires. In fact it needs a physical material medium to propagate and the electrical signals are confined within the cable or wire which transmits them. Typical forms of guided media include copper co-axial cables, fiber-optic cables and twisted-pair copper cables, which can be shielded or unshielded. Transmission of digital data through either guided or unguided communication involves the coding of the data at the sender's end, the modulation of the carrier signal, the demodulation of the signal on the receiving end and the decoding of the binary signal.

5.3.1   TWISTED PAIR

It is the most widely deployed media type across the world, as the last mile telephone link connecting every home with the local telephone exchange is made of twisted pair copper. It is also used as last mile connectivity to access the internet from home. They are also used in Ethernet LAN cables within homes and offices. They support low to High Data Rates (in order of Giga bits). However, they are effective only up to a maximum distance of a few kilometers, as the signal strength is lost significantly beyond this distance.

They come in two variants, namely UTP (unshielded twisted pair) and STP (shielded twisted pair). Within each variant, there are multiple sub-variants, based on the thickness of the material (like UTP-3, UTP-5, UTP-7 etc.). Twisted-pair cabling has insulated copper wires surrounded by an outer protective jacket. If the cable has an outer foil shielding, it is referred to as *shielded twisted pair (STP)*, which adds protection from radio frequency interference and electromagnetic interference. Twisted-pair cabling, which does not have this extra outer shielding, is called *unshielded twisted pair (UTP)*. Twisted-pair cable is cheaper and easier to work with.

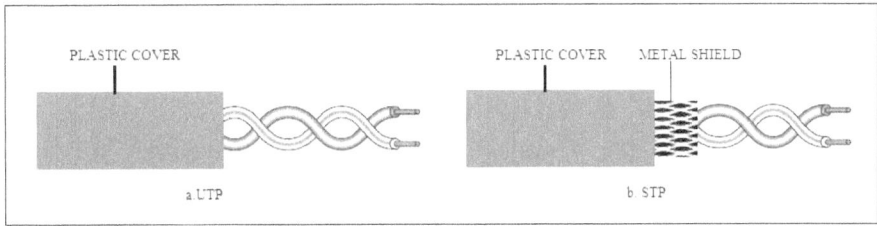

Figure 5.1   Twisted Pair Copper

The twisted-pair cable contains copper wires that twist around each other, as shown in Figure 5.1. This twisting of the wires protects the integrity and strength of the signals they carry. Each wire forms a balanced circuit, because the voltage in each pair uses the same amplitude, just with opposite phases. The tighter the twisting of the wires, the more resistant the cable is to interference and attenuation. UTP has several categories of cabling, each of which has its own unique characteristics. The twisting of the wires, the type of insulation used, the quality of the conductive material, and the shielding of the wire determine the rate at which data can be transmitted. The UTP ratings indicate which of these components were used when the cables were manufactured. Some types are more suitable and effective for specific uses and environments. The different UTP cable ratings are listed in Table 5.1.

Table 5.1   UTP cable ratings

UTP Category	Characteristics	Usage
Category 1	Voice grade telephone cable for up to 1 Mbps transmission rate.	Not recommended for network use, but modems can communicating over it.
Category 2	Data transmission up to 4 Mbps	Used in mainframe and minicomputer terminal connections, but not recommended for high-speed installations.
Category 3	10 Mbps for Ethernet and 4 Mbps for Token Ring	Used in 10 Base-T network installations.

Category 4	16 Mbps	Usually used in Token Ring networks.
Category 5	100 Mbps: has high twisting and thus low crosstalk.	Used in 100 Base-TX, CDDI, Ethernet, and ATM installations: most widely used in network installations.
Category 6	10 Gbps	Used in new network installations requiring high speed transmission. Standard for Gigbit Ethernet.
Category 7	10 Gbps	Used in new network installations requiring high speed transmission.

Copper cable has been around for many years. It is inexpensive and easy to use. A majority of the telephone systems today use copper cabling with the rating of voice grade. Twisted-pair wiring is the preferred network cabling, but it also has its drawbacks. Copper actually resists the flow of electrons, which causes a signal to degrade after it has traveled a certain distance. That is why cable lengths are recommended for copper cables; if these recommendations are not followed, a network could experience signal loss and data corruption. Copper also radiates energy, which means information can be monitored and captured by intruders. UTP is the least secure networking cable compared to coaxial and fiber. If a company requires higher speed, higher security, and cables to have longer runs which is beyond the permissible limit of copper cabling, fiber-optic cable may be a better choice.

5.3.2 CO-AXIAL

Co-axial copper cables have an inner copper conductor and an outer copper shield, separated by a di-electric insulating material, to prevent signal losses as shown in Figure 5.2. This is encased within a protective outer jacket. The term coaxial comes from the inner conductor and the outer shield sharing a geometric axis. Compared to twisted-pair cable, coaxial cable is more resistant to electromagnetic interference (EMI), provides a higher bandwidth (the bandwidth is 80 times more than twisted pair cable.), and supports the use of longer cable lengths.

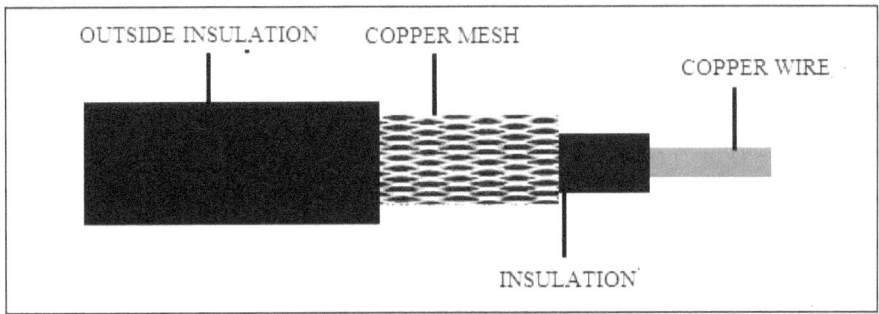

Figure 5.2   Co-axial copper cable

It is primarily used in cable TV networks and as trunk lines between telecommunication equipments.

1. It serves as an internet access line from the home.

2. It supports medium to High Data Rates.

3. It has much better immunity to noise and hence signal strength is retained for longer distances than in copper twisted pair media.

There are two main types of coaxial cables viz. thinnet and thicknet. Thinnet, also known as 10Base2, was commonly used to connect systems to backbone trunks of thicknet cabling. Thinnet can span distances of 189 meters and provide throughput up to 10 Mbps. Thicknet, also known as 10Base9, can span 900 meters and provide throughput up to 10 Mbps (megabits per second). The most common problems with coax cable are as follows:

1. Bending the coax cable past its maximum arc radius and thus breaking the center conductor.

2. Deploying the coax cable in a length greater than its maximum recommended length (which is 189 meters for 10Base2 or 900 meters for 10Base9).

3. Not properly terminating the ends of the coax cable with a 90 ohm resistor.

5.3.3 FIBER OPTICAL CABLES

Here, information is transmitted by propagation of optical signals (light) through fiber optic cables and not through electrical/electromagnetic signals. Due to this, fiber optics communication supports longer distances as there is no electrical interference. As the name indicates, fiber optic cables are made from glass (silica) and are as very thin as the human hair. They are coated with plastic also known as jacket.

As they support very high data rates, fiber optic lines are used as WAN backbone and trunk lines between data exchange equipments. They are also used for accessing internet from home through FTTH (Fiber-To-The-Home) lines. Additionally, they are used even for LAN environment with different LAN technologies like Fast Ethernet, Gigabit Ethernet etc. using optical links at the physical layer.

Fiber-optic cabling has higher transmission speeds that allow signals to travel over longer distances. Fiber cabling is not as affected by attenuation and EMI when compared to cabling that use copper. It does not radiate signals, as does UTP cabling, and is difficult to eavesdrop on; therefore, fiber-optic cabling is much more secure than UTP, STP, or coaxial.

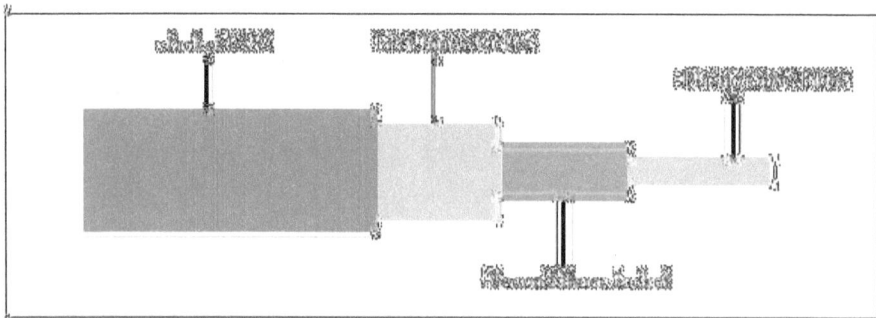

Figure 5.3   Fiber Optic Cable

Optical fiber consists of a core and a cladding layer, selected for total internal reflection due to the difference in the refractive index between the two which is shown in Figure 5.3. The fiber cables which are presently using, the cladding is usually coated with a layer of acrylate polymer or polyimide. This coating protects the fiber from damage but does not contribute to its optical waveguide properties. Individual coated fibers (or fibers formed into ribbons or bundles) then have a tough resin buffer layer and/or core tube(s) extruded around them to form the cable core. Several layers of protective sheathing, depending on the application, are added to form the cable. Rigid fiber assemblies sometimes put light-absorbing (dark) glass between the fibers, to prevent light that leaks out of one fiber from entering another. This reduces cross-talk between the fibers, or reduces flare in fiber bundle imaging applications. For indoor applications, the jacketed fiber is generally enclosed, with a bundle of flexible fibrous polymer *strength members* like aramid (e.g. Twaron or Kevlar), in a lightweight plastic cover to form a simple cable. Each end of the cable may be terminated with a specialized optical fiber connector which allow it to be easily connected and disconnected from transmitting and receiving equipment.

For use in more strenuous environments, a much more robust cable construction is required. In *loose-tube construction* the fiber is laid helically into semi-rigid tubes, allowing the cable to stretch without stretching the fiber itself. This protects the fiber from tension during laying and due to temperature changes. Loose-tube fiber may be *dry block* or gel-filled. Dry block offers less protection to the fibers than gel-filled, but costs considerably less. Instead of a loose tube, the fiber may be embedded in a heavy polymer jacket, commonly called tight buffer construction. Tight buffer cables are offered for a variety of applications, but the two most common are *Breakout* and *Distribution*. Breakout cables normally contain a ripcord, two non-conductive dielectric strengthening members (normally a glass rod epoxy), an aramid yarn, and 3 mm buffer tubing with an additional layer of Kevlar surrounding each fiber. The ripcord is a parallel cord of strong yarn that is situated under the jacket(s) of the cable for jacket removal. Distribution cables have an overall Kevlar wrapping, a ripcord, and a 900 micrometer buffer coating surrounding each fiber. These *fiber units* are commonly bundled with additional steel strength members, again with a helical twist to allow stretching.

A critical concern in outdoor cabling is to protect the fiber from contamination by water. This is accomplished by the use of solid barriers such as copper tubes, and water-repellent jelly or water-absorbing powder surrounding the fiber. Finally, the cable may be armored to protect it from environmental hazards, such as construction work or gnawing animals. Undersea cables are more heavily armored in their near-shore portions to protect them from boat anchors, fishing gear, and even sharks, which may be attracted to the electrical power that is carried to power amplifiers or repeaters in the cable. Modern cables come in a wide variety of sheathings and armor, designed for applications such as direct burial in trenches, dual use as power lines, installation in conduit, lashing to aerial telephone poles, submarine installation, and insertion in paved streets.

5.3.4   CABLING CONSIDERATIONS

Cables are extremely important within networks, and when they experience problems, the whole network can be complex and problematic. This section addresses some of the common cabling issues that many networks experience.

5.3.4.1   Noise

Noise on a line is usually caused from the adjoining devices or from the environment. Noise can be produced by motors, computers, copy machines, fluorescent lighting, and microwave ovens. This background noise can combine with the data being transmitted over the cable and distort the signal.

5.3.4.2   Cabling Connection Types

Cables follow standards, for interoperability and connectivity between common devices and environments. The standards are developed and maintained by the Telecommunications Industry Association (TIA) and the Electronic Industries Association (EIA). The TIA/EIA standards enable the design and implementation of structured cabling systems for commercial buildings. The majority of the standards define cabling types, distances, connections, cable system architectures, cable termination standards and performance characteristics, cable installation requirements, and methods of testing installed cable. The following are commonly used physical interface connection standards.

1. RJ-11 is used for terminating telephone wires,

2. RJ-45 is used for terminating twisted-pair cables in Ethernet LAN, and

3. BNC (British Naval Connector) is used for terminating coaxial cables.

5.3.4.3   Attenuation

Attenuation is the loss of signal strength while it propagates through the medium. The attenuation is more for a longer cable which causes the data signal to deteriorate. That is why standards suggest cable-run lengths. The effects of attenuation increases with higher frequencies. This means that cable used to transmit data at higher frequencies should have shorter cable runs to ensure attenuation does not become an issue.

Basically, the data are in the form of electrons, and these electrons have to swim through a copper wire. However, this is more like swimming upstream, because there is a lot of resistance on the electrons working in this media. After a certain distance, the electrons start to slow down and their encoding format loses form. If the form gets too degraded, the receiving system cannot interpret them any longer. If a network administrator needs to run a cable longer than its recommended segment length, he needs to insert a repeater that will amplify the signal and ensure it gets to its destination in the right encoding format. Attenuation can also be caused by cable breaks and malfunctions. If a cable is suspected of attenuation problems, cable testers can inject signals into the cable and confirm the fault.

5.3.4.4   Crosstalk

Crosstalk is a phenomenon caused by the electric or magnetic fields of one communication channel spill over to the signals of adjacent channel. When the different electrical signals mix, the integrity of the signals degrades and data may get corrupted. In a telephone circuit, crosstalk can result in disturbing the hearing part of a voice conversation. It can occur in microcircuits within computers, audio equipment and within network circuits. UTP is much more vulnerable to crosstalk than STP or coaxial, because it does not have extra layers of shielding to protect against crosstalk.

5.3.4.5   Fire-rated and Flame-retardant Cables

Fire-resistive or fire-rated cable is a cable that will continue to operate in the presence of a fire. This is commonly known as circuit integrity

(CI) cable. Certain cables , when on fire produces hazardous gases that would spread throughout the building quickly. Network cabling that is placed in these types of areas, referred as plenum space, must meet a specific fire rating to ensure it will not produce and release harmful chemicals in case of a fire.

Flame-retardant cable is a cable that will not convey or propagate a flame as defined by the flame-retardant or propagation tests. Flame-retardant tests measure flame propagation for both horizontal and vertical applications. There are also plenum cable flame tests when it is used in ducts, plenums or other spaces.

A flame-retardant cable is not a fire-rated cable. There are certain but specific differences between flame-retardant cables and fire-resistive cables. Flame-retardant cables resist the spread of fire into a new area, whereas fire-resistive cables maintain circuit integrity and continue to work for a specific time under defined conditions. These circuit integrity cables continue to operate in the presence of a fire and are sometimes called 1-hour or 2-hour fire-rated cables. CI cables are needed when it is most essential and critical for life safety or to prevent a plant shut down.

Depending upon the situation, while setting up a network, it is important to select the appropriate types of wire. Cables should be installed in unexposed areas so they are not easily tripped over, damaged, or eavesdropped upon. The cables should be strung behind walls and in the protected spaces as in dropped ceiling. In environments that require extensive security, wires are encapsulated within pressurized conduits so if someone attempts to access a wire, the pressure of the conduit will change, causing an alarm to sound and a message to be sent to the security men.

Note: A plenum is the air return path of a central air handling system. It can be either ductwork or open space over a suspended ceiling or raised floor.

5.4   UNGUIDED MEDIA

Unguided media are more commonly known as wireless media or unbounded media, in which data is transferred into electromagnetic waves and sent through free space without guiding any specific direction. Hence the name unguided media and are classified as wireless transmission which is a quickly expanding field of technology for networking. These are microwave, cellular radio, radio broadcast and satellite. As wireless technologies continue to proliferate, the organization's security efforts must go beyond locking down its local network. Security should be an end-to-end solution that addresses all forms, methods, and techniques of communication.

Different types of unguided communication are classified based on the frequency spectrum used for communication, the distance between the end stations and the type of encoding used for the communication. Broadband wireless signals occupy frequency bands that may be shared with microwave, satellite, radar, and ham radio use. Unguided communication allows electromagnetic signals to travel between antennas, some of which are on satellites. Antennas can provide point-to-point communication or can send their signals in all directions. These technologies are used for television transmissions, cellular phones, satellite transmissions, spying, surveillance, and garage door openers. The different unguided medias which are employed in distributed SCADA system are briefly explained below.

5.4.1 MICROWAVES COMMUNICATION

In this kind the data is transferred via air. The waves travel in a straight line. The data is received and transferred via microwave stations. The speed at which data is transferred is 190 Mbps. The two main microwave wireless transmission technologies are satellite (ground to orbiter to ground) and terrestrial (ground to ground). They are widely used by telephone and cable companies.

5.4.2 TERRESTRIAL COMMUNICATION

This type of communication is limited to line-of-sight (LOS) transmission. This means that microwaves are transmitted in a straight line and that no obstructions can exist, such as buildings or mountains, between microwave stations as shown in Figure 5.4. To avoid possible obstructions, microwave antennas often are positioned on the tops of buildings, towers, or mountains. It finds applications in long-distance telecommunication service. It requires fewer amplifiers or repeaters than coaxial cable but it is line-of-sight transmission. Short point-to-point links, Data link between local area network, closed-circuit TV, etc.

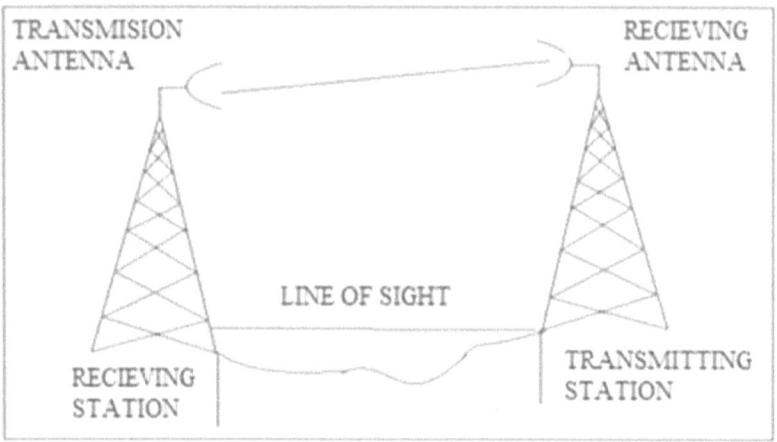

Figure 5.4   Terrestrial communication

5.4.3   SATELLITE COMMUNICATION

The satellites are located at a distance of 22300 miles above the earth as shown in Figure 5.5. The signals are received from earth stations. Devices like GPS and PDAs also receive signals from these earth based stations. The process of transferring and receiving data takes place within few seconds. The data is transferred at a speed of 1 Gbps. They are used for purposes like weather forecast, military communication, radio transmission, satellite TV, data transmission, etc.

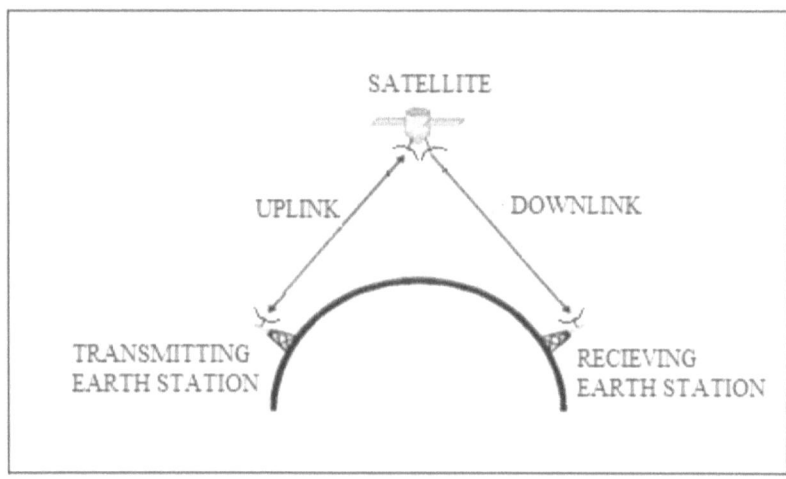

Figure 5.5   Satellite communication

Today the cost of satellite launching has brought down significantly with advanced and reusable launching modules. This made extensive use of satellite communication to provide wireless connectivity between different locations. But for two different locations to communicate via satellite links, they must be brought within the satellite's line-of-sight and area covered by the satellite called footprint. The information or data is appropriately modulated by the ground station and is transmitted to the satellite. A transponder on the satellite receives this signal, amplifies it, and relays it to the receiver. The receiver at the ground station must have an antenna, usually a circular dish like structure normally placed on top of buildings. The antenna contains one or more microwave receivers, depending upon the number of satellites it is accepting data from. Satellites provide broadband transmission. If a user is receiving radio or TV data, then the transmission is set up as a one-directional network. If a user is using this connection for internet, then the transmission is set up as a bi-directional network. The available bandwidth depends upon the antenna, terminal type and the service rendered by the service provider.

As the satellites are kilometers above the Earth, time-critical applications may experience delay as the signal has to traverse to and from the satellite. Hence these types of satellites are placed into a low Earth orbit, keeping the distance between the ground stations and the satellites is less when compared to other types of satellites. Further, smaller receivers can be used for reception of signals, which makes low-Earth-orbit satellites ideal for cellular communication, radio and TV stations, Internet, etc.

5.4.4 MOBILE COMMUNICATION

High frequency Radio Frequency (RF) waves are used for the transmission of data in mobile communication. Here one can receive and make calls and also access the internet. Earlier, when Gopi wanted to call Jessy, they had to depend on the telephone which had a physical connection. The telephone would only reach as far as the telephone cord that it was attached to, and the handset was also physically connected. So they actually had to sit in one place to carry out a conversation. This model worked for almost 100 years, but once mobile phones were materialized and available to everyone at a low price, there was no going back. Today mobile wireless communication has exploded with its popularity and capabilities.

Mobile phone is a device that can send voice and data over wireless radio links. It connects to a cellular network, which is connected to the Public-

Switched Telephone Network (PSTN). So instead of a physical cord and connection that connects the phone and the PSTN, one should have a device which connects indirectly to the PSTN. Radio is the transmission of signals via electromagnetic waves within a certain frequency range. A cellular network distributes radio signals over delineated areas, called cells. Each cell has at least one fixed-location transceiver at base station, and is joined to other cells to provide connections over large geographic areas. So as somebody is talking on his mobile phone and he moves out of range of one cell, the base station in the original cell sends his connection information to the next base station so that his call is not dropped and he can continue his conversation.

This switching from one cell to another cell, does not arise the requirements of an infinite number of frequencies to work with. Lots of people around the world may be using their cell phones simultaneously. All of these calls take place with a set of frequencies. An elementary representation of a cellular network is shown in Figure 5.6.

Individual cells can use the same frequency range, as long as they are not next to each other. So the same frequency range can be used in every other cell, which drastically decreases the amount of ranges required to support simultaneous connections. The communication service providers have to come up with different ways to allow millions of consumers to use this finite frequency range in a flexible manner.

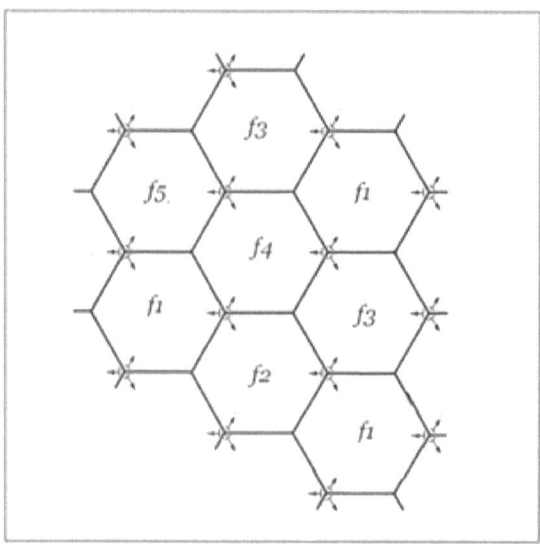

Figure 5.6   Non-adjoint cells can use the same frequency ranges

Presently, mobile wireless communication has been gradually made up of complex and powerful multiple access technologies, which are mentioned below.

1. Frequency division multiple access (FDMA)

2. Time division multiple access (TDMA)

3. Code division multiple access (CDMA)

4. Orthogonal frequency division multiple access (OFDMA)

The characteristics of each of these technologies are briefly described below as they are the foundational constructs of the various cellular network generations.

When the useful bandwidth of the medium exceeds the required bandwidth of a signal, Frequency division multiple access (FDMA) is a better solution. Here a number of signals can be carried simultaneously by modulating the signals into different carrier frequency with sufficient separation between them. In 3G FDMA, the available frequency range is divided into sub-bands called channels, and one channel is assigned to each cell phone. The subscriber has exclusive use of that channel while the call is made, or until the call is terminated., no other calls or conversations can be made on that channel during that call. Using FDMA, multiple users can share the frequency range without the risk of interference between the simultaneous calls. FDMA was used in 1G.

When the achievable bit rate of the medium, exceeds the required data rate of a digital signal, Time Division Multiple Access (TDMA) is a better solution. TDMA increases the speed and efficiency of the cellular network by taking the RF spectrum channels and dividing them into time slots. In TDMA systems, time is divided into frames and each frame is divided into slots. TDMA requires information regarding starting and ending time of each slot which is known to both the source and the destination. Mobile communication systems such as Global System for Mobile Communication (GSM), Digital AMPS (D-AMPS), and Personal Digital Cellular (PDC) use TDMA.

As the term implies, CDMA is a form of multiplexing, which allows numerous signals to occupy a single transmission channel, optimizing the use of available bandwidth. It assigns a unique code to each voice call or data transmission to uniquely identify it from all other transmissions sent over the cellular network. In a CDMA spread spectrum network, calls are spread throughout the entire frequency band. CDMA permits all users of the network to simultaneously use every channel in the network. At the same time, a particular cell can simultaneously interact with multiple other cells. These features make CDMA a very powerful technology, for the mobile cellular networks that presently dominate the wireless space. It has improved voice and data communication capability and is more secure.

Orthogonal Frequency Division Multiple Access (OFDMA) is a combination of FDMA and TDMA. In former implementations of FDMA, the different frequencies for each channel were widely spaced to allow analog hardware to separate the different channels. In OFDMA, each of the channels is subdivided into a set of closely spaced orthogonal frequencies with narrow bandwidths (subchannels). Each of the different subchannels can be transmitted and received simultaneously in a Multiple Input and Output (MIMO) manner. The use of orthogonal frequencies and MIMO allows signal processing techniques to reduce the impacts of any interference between different subchannels and to correct for channel impairments, such as noise and selective frequency fading. Mobile wireless technologies have gone through a whirlwind of confusing generations. The first generation (1G) dealt with analog transmissions of voice-only data over circuit-switched networks. This generation provides a throughput of around 19.2 Kbps. The second generation (2G) allows for digitally encoded voice and data to be transmitted between wireless devices, like cell phones.

The third-generation (3G) networks became available by incorporating FDMA, TDMA, and CDMA. In 3G, circuit switching is replaced by packet switching. The flexibility to support a great variety of applications and services are the main advantages of 3G. The modular design allows expandability, backward compatibility with 2G networks, interoperability among mobile systems, global roaming, and Internet services. 3G with more enhancements, referred to as 3.9G or mobile broadband, is taking place under the title of Third Generation Partnership Project (3GPP). 3GPP has a number of new technologies such as Enhanced Data Rates for GSM Evolution (EDGE), High-Speed Downlink Packet Access (HSDPA), CDMA2000, and Worldwide Interoperability for Microwave Access (WiMax). There are two competing technologies that fall under the umbrella of 4G, which are Mobile WiMax and Long-Term Evolution

(LTE). A 4G system does not support traditional circuit-switched telephony service as 3G does, but works over a purely packet based network. 4G devices are IP-based and are based upon OFDMA.

Communication engineers and scientists are actively involved in developing Fifth-Generation (5G) mobile communication, but standards requirements and implementation are not expected in near feature. Each of the different mobile communication generations has taken advantage of the improvement of hardware technology and processing power. The increase in hardware has allowed for more complicated data transmission between customers as more customers want to use mobile communications.

Table 5.2 illustrates some of the main features of the 1G through 4G networks. It is important to note that this table does not and cannot easily cover all the aspects of each generation. Earlier generations of mobile communication have considerable variability between countries. The variability was due to country-sponsored efforts before agreed-upon international standards were established. Various efforts between the ITU and countries have attempted to minimize the differences.

Table 5.2   Different characteristics of Mobile Technology

	1G	2G	3G	4G
Spectrum	900MHz	1800MHz	2GHz	Various
Multiplxing Used	FDMA	TDMA	CDMA	OFDMA
Voice Support	Basic Telephony	Caller Id and Voice Mail	Conference Calls and Low Quality Video	High Definition Video
Messaging Features	Not available	Text Only	Graphic and Formatted Text	Full Unified Messaging
Data Support	Not available	Circuit Switched	Packet Switched	Native IPv6
Target Data Rate	Not available	119-128Kbps	2 Mbps (10Mbps in 3.9G)	100Mbps 1Gbps having mobility

5.5   SCADA COMMUNICATION TECHNOLOGIES

SCADA communication technologies broadly classified into two viz. wired (guided) or wireless (unguided).

5.5.1   WIRED OR GUIDED MEDIA TECHNOLOGIES

Wired or guided media technologies used in industrial SCADA and power system SCADA are Copper UTP, Coaxial, Optical Fiber, Power Line Carrier Communication (PLCC), Broadband over Power line (BPL) and HomePlug. A brief description of these technologies is given below.

5.5.1.1   Copper UTP

Unshielded Twisted Pair (UTP) is a popular type of cable that consists of two unshielded wires twisted around each other which has been explained earlier in this chapter. Low cost makes it attractive, and is used extensively for local-area networks (LANs) and telephone connections. UTP cabling does not offer as high bandwidth or good protection from interference when compared to coaxial or fiber optic cables, but easier to work with.

5.5.1.2   Optical Fiber

Fiber cables offer several advantages over traditional long-distance copper cabling. The bandwidth for an optical fiber which has the same thickness as that of a copper cable is much higher. Fiber cables rated at 10 Gbps, 40 Gbps and even 100 Gbps are standard. As light can travel much longer distances in a fiber cable without attenuation, the need for signal boosters are less. Further fiber is less susceptible to interference. Traditional copper network cable requires shielding to protect it from electromagnetic interference. Though shielding helps to prevent interference, it is not always sufficient especially, when many cables are strung together in close proximity to each other. But fiber optic cables which are made up of glass or silica are capable of avoiding most of these problems.

5.5.1.3   Fiber to the home (FTTH)

Fiber to the Home (FTTH) broadband is one of the fine solutions for providing connectivity and presently it is generally referred as fiber optic cable connections to individual houses. It can deliver a large number of digital information including voice, video, and data with more efficiently and economically than coaxial cable. FTTH depend on both active and passive optical networks to provide connectivity. FTTH technology is supposed to be an apt standard which can provide connectivity without much web traffic congestion.

5.5.1.4   Hybrid Fiber Coax (HFC)

A hybrid fiber coaxial (HFC) network is a guided media technology where optical fiber cable and coaxial cable are used in different portions of a communication network to carry broadband information such as video, data, and voice. Using HFC, service providers install fiber optic cable as the backbone from the cable distribution center to the serving nodes located close to customers. Then using coaxial cables, connections are provided to the customers by connecting to the nodes of the fiber optic backbone. The main advantage of using HFC as backbone is that some of the features of fiber optic cable such as high bandwidth, improved noise immunity and interference susceptibility, etc. can be brought close to the customer premises without the replacement of the existing coaxial cable if it exists already.

5.5.1.5   Power Line Carrier Communication (PLCC)

Power-line carrier communication (PLCC) is a communication method that uses electrical wiring to simultaneously carry both data and electric power. Although many power utilities use this for long distance to send data, it is very rarely used within the buildings. But recently PLCC is used to achieve load shedding in Advanced Metering Infrastructure (AMI). As it uses the existing power lines, there is no additional investment on cables and structural alterations to the building. It is an economical and a reliable technique to achieve bi-directional communications which is one of the prime requirements in Smart Grid.

Different PLCC technologies are required for different applications, ranging from home automation to Internet access which is often called broadband over power lines (BPL). Most PLCC technologies limit themselves to one type of wire but some can cross between the distribution network and customer premises wiring. Typically transformers prevent propagating the signal, which requires multiple technologies to form very large networks. Various data rates and frequencies are used in different situations. But the main drawback is low bandwidth and point-to-point communication. Although long distance communication is possible, it poses significant challenges, especially in developing countries where the disturbances on transmission lines cause issues.

5.5.1.6   Broadband over Power Line (BPL)

PLCC can be broadly grouped as narrow band PLCC and broadband PLCC, also known as low frequency and high frequency respectively. There are four basic forms of PLCC and they are:

1. Narrowband internal applications where home wiring is used for home automation and intercoms but with low bit rate,

2. Narrowband outdoor applications which are mainly used by the power distribution utilities for AMR,

3. Broadband in-house mains power wiring can be used for high speed data transmission for home networking, and Broadband over Power Line which uses outdoor mains power wiring and can be used to provide broadband internet.

Broadband PLCC works at higher frequencies. High data rates (up to 100s of Mbps) are used in shorter-range applications. In fact BPL is a system to transmit two-way data over existing AC medium voltage electrical distribution wiring, between transformers, and AC low voltage wiring between transformer and customer outlets. This makes it suitable for indoor as well as outdoor applications. This avoids the expense of a dedicated network of wires for data communication, and the expense of maintaining a dedicated network of antennas, radios and routers in wireless network.

5.5.1.7   HomePlug

HomePlug is a Power line networking which uses power line communications. The specification of Homeplug is defined by the home networking technology that connects devices to each other through the power lines within a home. HomePlug certified products connect PCs and other devices that use Ethernet, USB and 802.11 Wi-Fi technologies to the power line via a HomePlug bridge or adapter. Some products have HomePlug technology built-in. It is one of the cheapest forms of home networking and has a low start-up cost and minimal IT workload. Also it will not have an adverse effect on home electric bills.

As consumers need to connect more devices, the need for high throughput connectivity has tremendously amplified. HomePlug technology enables electrical wires of homes to distribute broadband Internet, Ultra High Definition video streaming, virtual reality, digital music and smart energy applications. HomePlug hybrid networking products are used by consumers and service providers worldwide to provide both wired home networking connectivity and Wi-Fi extension throughout the home in dead spots and areas furthest from the router. Growing smart home trends will also continue to increase the strain on home networks and the need for a singular network for both entertainment and IoT products.

HomePlug adapters are available for advertising physical rates of 200Mbps, 900Mbps and 1Gbps. This claims more bandwidth than an ordinary broadband connection, and perfect for streaming HD videos, downloading large files and online gaming. One benefit of this high capacity is the ability for multiple simultaneous communication streams. In SCADA, PLCC is mainly used in PSS for tele-protection and tele-monitoring between electrical substations through power lines at high voltages, such as 110 kV, 220 kV, and 400 kV. But this can also be used by utilities for Energy Management Systems (EMS), fraud detection and network management, Advanced Metering Infrastructure (AMI), Demand Side Management (DSM), load control, and demand response (DR) etc.

5.5.2 WIRELESS OR UNGUIDED MEDIA TECHNOLOGIES

Presently, the unguided or wireless technology has become the most thrilling area in communications and M2M networking. The advancement of wireless communication has revolutionized Industrial SCADA networks as well. Some of the wireless technologies which find applications in DCS and Smart Grid are Frequency Hopping Spread Spectrum (FHSS), 3G Cellular, Wi-Fi, WiMax, ZigBee, ZWave, and VSAT which are briefly described below.

5.5.2.1   IEEE and Wireless Standards

Standards are developed so that many different vendors can create various products that will work together seamlessly. Standards are usually developed on a consensus basis among the different vendors in a specific industry. The Institute of Electrical and Electronics Engineers (IEEE) develops standards for a wide range of technologies and wireless being one of them.

The 802.11 standard outlines how wireless clients and Access Points communicate, lays out the specifications of their interface, dictates how signal transmission should take place, and describes how authentication, association, and security should be implemented. IEEE created several task groups to work on specific areas within wireless communications. Each group had its own focus and was required to investigate and develop standards for its specific section. The letter suffixes indicate the order in which they were proposed and accepted. As an example, one of the members of the 802.11 family which specifies Wi-Fi is succinctly described below.

802.11b High Rate or Wi-Fi is an extension to 802.11 which applies to wireless LANS and provides 11 Mbps transmission in the 2.4 GHz band. 802.11b uses only Direct Sequence Spread Spectrum (DSSS). 802.11b is ratification to the original 802.11 standard, allowing wireless functionality comparable to Ethernet.

Wireless LAN products are being developed following the stipulations of this 802.11i wireless standard. Customers should review the certification issued by the Wi-Fi Alliance before buying wireless products as it assesses the systems against the 802.11i proposed standard.

5.5.2.2   Frequency Hopping Spread Spectrum (FHSS)

Frequency-hopping spread spectrum (FHSS) transmission is the repeated switching of frequencies during radio transmission to reduce interference and avoid interception. It is one of the better methods to counter eavesdropping, and to block jamming of telecommunications. It also has the capability to minimize the effects of unintentional interference. In FHSS, the transmitter hops between available narrowband frequencies within a specified broad channel in a pseudo-random sequence known to both sender and receiver. A short burst of data is transmitted on the current narrowband channel, and then transmitter and receiver tune to the next frequency in the sequence for the next burst of data. In most systems, the transmitter will hop to a new frequency more than twice per second. As no channel is used for a long time, and the possibility of any other transmitter being in the same channel at the same time are low, FHSS is often used as a method to allow multiple transmitter and receiver pairs to operate in the same space, in the same channel, at the same time. Direct Sequence Spread Spectrum (DSSS) is a related technique. It also spreads a signal across a wide channel, but it does so, all at once, instead of in discrete bursts separated by hops. The spreading code has a higher chip rate, which results in a wideband time continuous scrambled signal. DSSS has certain advantages such as having the best discrimination against multipath signals, capability of effectively avoiding intentional interference like jamming, and better noise immunity than FHSS system.

The frequency hopping spread spectrum over a fixed-frequency transmission method has following primary advantages.

1. The method is very resistant to narrow band interference since the spread signal causes the interfering signal to recede into the background.

2. The signals are very difficult to intercept. FHSS signals seem like there has been an increase in background noise when a narrow band receiver detects them. In order to intercept the signal, the pseudorandom transmission hopping sequence has to be known.

3. FHSS transmissions can share frequency bands with a number of other types of conventional transmissions without causing significant interference. Each of these signals causes minimal interference and allows the bandwidth to be used more effectively.

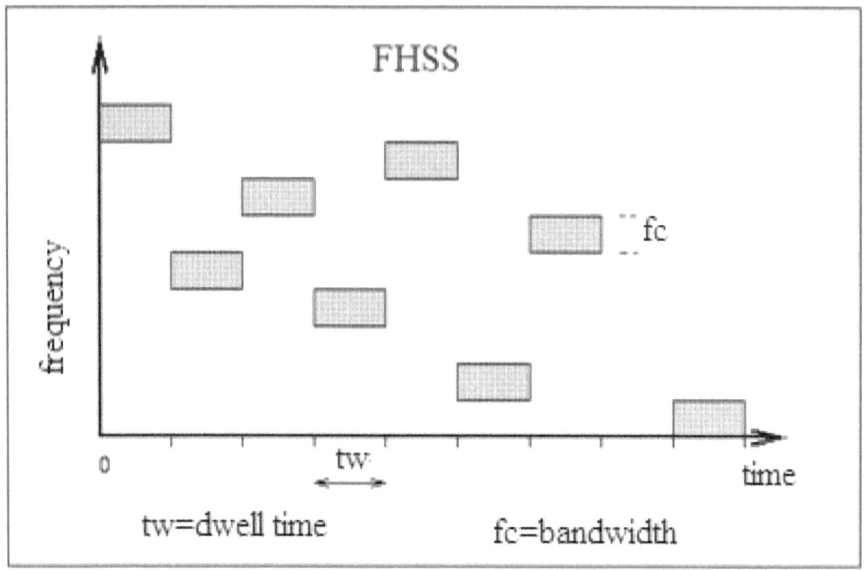

Figure 5.7   changing carrier frequency at random

However FHSS faces certain disadvantages as it requires greater bandwidth than when a single carrier frequency is used. Also, one of the most problematic aspects of this technology is to initially synchronize the transmitter and receiver. Further significant time is required to master the process of synchronizing the transmitter and receiver.

5.5.2.3   3G Cellular

3G also follows the same pattern of G's which has been introduced in the early 1990's by the ITU. The pattern is actually a wireless initiative called the International Mobile Communications (2000IMT-2000). 3G therefore comes just after 2G and 2.9G, the second generation technologies. 2G technologies include, the Global System for Mobile (GSM). 2.9G brings standards that are midway between 2G and 3G, including the General Packet Radio Service (GPRS), Enhanced Data rates for GSM Evolution (EDGE), Universal Mobile Telecommunications System (UMTS), etc. A characteristic Comparison of the various Gs is given in Table 5.3.

Table 5.3 Characteristic Comparison of the various Gs

2G	3G	3.9G (3GPP)	4G
Higher bandwidth than 2G	voice and data are integrated	Higher data rates	IP packet switched network
Always on technology for e-mail	Packet switched technology,	Use of OFDMA technology.	Packet switched technology.

This is a low cost solution to enable long-range communication both within the plant and from field to the MCC. It can be rolled out quickly using the existing cellular infrastructure. Most smart meters in future are expected to use this technology to communicate with the Meter Data Management Systems (MDMS). The constraint is that it can become undependable if natural disaster strikes.

In DCS like Smart Grid, for remote monitoring and control solution, network coverage for a broad range of locations is necessary. In order to achieve coverage, the remote devices deployed are fitted with wireless modems that enable communication on a range of frequencies and protocols. This allows selection of either 2G EDGE, 3G HSPDPA, with a flexible architecture to accommodate Long Term Evolution (LTE) when deployed by carriers.

5.5.2.4 Wi-Fi

Wi-Fi is based on IEEE 802.11 standard, a trademark for the open standards, is a popular Radio Frequency (RF) wireless networking technology to provide high-speed Internet and network connections. One can attain bandwidth from 9 to 94Mbps. The limited distance of 100 to 290m and poor reception in buildings are the main disadvantages. Wi-Fi networks have no physical wired connection between sender and receiver as it uses Radio Frequency technology. RF being a member of the electromagnetic spectrum associated with radio wave propagation (2.9GHz for 802.11b or 11g, 9GHz for 802.11a), the The RF current supplied to an antenna creates an electromagnetic field which makes radio waves to propagate through space. The keystone of any wireless network is an Access Point (AP). The primary job of an AP is to broadcast a wireless signal that computers can detect and tune into. In order to connect to an AP and join a wireless network, computers and devices must be equipped with wireless network adapters.

Wi-Fi Alliance: Wi-Fi is supported by many applications and devices which include mobile phones, video game consoles, home area networks, PDAs, major operating systems, and many consumer electronic products. Any products that are tested and approved as *Wi-Fi Certified* by the Wi-Fi Alliance are interoperable to each other, even if they are from different make and model. This implies, a consumer with a Wi-Fi Certified product can use any brand of access point with any other brand of client hardware which is also Wi-Fi Certified. Products that pass this certification are required to carry an identifying seal on their packaging that states Wi-Fi Certified and indicates the radio frequency band used.

5.5.2.5   WiMax

WiMax stands for Worldwide Interoperability for Microwave Access, follows IEEE 802.16d communication standard. One can get up to 70 Mbps bandwidth over 70 to 80 kilometers. This bandwidth is a savior when an environmental disaster strikes a densely populated area and all smart meters start communicating the outages at same time. WiMax can handle this increased traffic and because of this, WiMax can be an ideal backhaul medium for in-premise Wi-Fi and ZigBee devices. Higher cost and poor market adoption are major constraints at the moment and is not a replacement for Wi-Fi or wireless hotspot technologies. However, all-in-all, it can be cheaper to implement WiMax instead of standard wired hardware like DSL.

WiMax equipment exists in two basic forms viz. base stations, installed by service providers to deploy the technology in a coverage area and receivers, installed in clients. WiMax is developed by an industry consortium, overseen by a group called the WiMax Forum, who certifies WiMax equipment to ensure that it meets technology specifications. Its technology is based on the IEEE 802.16 set of wide-area communications standards. Presently deployments of WiMax are in fixed locations, but development of mobile version is underway.

WiMax can be installed faster than other internet technologies because it can use shorter towers and less cabling, supporting even non-line-of-sight (NLoS) coverage across an entire city or country. WiMax isn't just for fixed connections either, like at home. One can also subscribe to a WiMax service for his mobile devices since USB dongles, laptops and phones can have the technology built-in. In addition to internet access, WiMax can provide voice and video transferring capabilities as well as telephone access. Since WiMax transmitters can span a distance of several miles with data rates reaching up to 30–40 Megabits Per Second (Mbps), it's easy to see its advantages, especially in areas where wired internet is impossible or too costly to implement.

However it has certain disadvantages as WiMax is wireless by nature, and they are briefed below.

1. WiMAX technology offering long distance data range which is 70 kilometer and high bit rate which is 70Mbit/s. But both features don't work together. Increasing the distance range will decrease the bit rate and increasing bit rate will reduce the distance range,

2. Any user closer to the tower can get high speed while user at the cell edge from the tower gets much diminished speed of connectivity,

3. Functionality could go down with sharing of bandwidth, if more than one user exists in a single sector and a range of 2 to 8 or 12 Mbps services cannot be ensured, then additional radio cards are added to the base station to boost the capability,

4. Susceptible to weather conditions like rain which could interrupt the signal and the wireless equipment could cause interference, and

5. WiMAX is a very power intensive technology and requires strong electrical support.

5.5.2.6 ZigBee

It is a low cost and low power technology and uses unlicensed spectrum which covers only short distance. ZigBee is gaining popular in the home energy market and multiple ZigBee enabled products are available in the market. ZigBee enables the smart meter to communicate with home appliances which helps to shed the load. ZigBee technology enables the coordination of communication among thousands of tiny sensors, which can be scattered throughout the offices, farms, factories, picking up information about the operations and process. They are designed to consume very little energy because it will be left in place for five to ten years and their batteries need to last. ZigBee devices communicate efficiently, passing data over radio waves from one to the other like a human chain. At the end of the line, the data can be dropped into a computer for analysis or picked up by another wireless technology like Wi-Fi or WiMax.

ZigBee is constituted of mesh technology and this makes it more robust. ZigBee alliance has come up to ensure interoperability among home appliances but the limited distance and inability to penetrate concrete walls are major constraints. Further the low power consumption limits transmission distances to 10–100 meters line-of-sight, depending on power output and environmental characteristics. ZigBee devices can transmit data over long distances by passing data through a mesh network of intermediate devices to reach more distant ones. ZigBee is typically used in low data rate applications that require long battery life and secure networking. ZigBee networks are secured by 128 bit symmetric encryption keys. ZigBee has a defined rate of 290 kbit/s, best suited for intermittent data transmissions from a sensor or input device. ZigBee is an ideal solution for Personnel Area Netwoks. However, ZigBee is not for situations with high mobility among nodes. Hence, it is not suitable for tactical ad hoc radio networks in the battlefield, where high data rate and high mobility is present and needed.

5.5.2.7 ZWave

ZWave is a low cost, low power RF signaling and control protocol used for communication among devices preferably in home area networks. It is developed by a private company named Zensys, Inc. a start-up company based in Denmark. The technology of ZWave is based on the concepts of ZigBee. However ZWave attempts to build simpler and less expensive devices than ZigBee. Presently, Zensys is acquired by Sigma Designs of Milpitas, CA. Many manufacturers make ZWave compatible products, mostly for the building automation and HVAC. ZWave operates at 908.42 MHz in the US and 868.42 MHz in Europe using a mesh networking topology. A ZWave network can contain up to 232 nodes, although reports exist of trouble with networks containing over 30–40 nodes. ZWave operates using a number of profiles, but the manufacturer claims they interoperate. One should be careful when selecting products as some products from certain manufacturers are not compatible with other manufacturers' products.

ZWave uses GFSK modulation with Manchester channel encoding. A network controller, at the center monitors and controls the ZWave network. Each product in the home must be included to the ZWave network before it can be controlled via ZWave. Each ZWave network is identified by a Network ID and each device is further identified by a Node ID. The Network ID is the common identification of all nodes belonging to one logical ZWave network. Network ID has a length of four bytes and is assigned to each device by the primary controller when the device is added into the network. Nodes with different Network IDs cannot communicate with each other. The Node ID is the address of the device/node existing within network. The Node ID has a length of one byte. ZWave uses a source-routed mesh network topology and has central controller. Devices can communicate to one another by using intermediate nodes to route around and circumvent household obstacles or radio dead spots that might occur though a message called healing. A ZWave network can consist of up to 232 devices with the option of bridging networks if more devices are required.

ZWave Alliance: The ZWave Alliance was established as a consortium of companies that make connected appliances controlled through apps on smartphones, tablets or computers using ZWave wireless mesh networking technology. The alliance is a formal association focused on both the expansion of ZWave and the continued interoperability of any device that utilizes ZWave.

ZWave Security: Recently, ZWave Alliance announced stronger security standards for devices receiving ZWave Certification. The security standards are known as Security 2 (S2), it provides advanced security for devices to be connected to ZWave gateways and hubs. Encryption standards for transmissions between nodes, and new pairing procedures for each device, with unique PIN for each devices are established. The new layer of authentication is intended to prevent hackers from taking control of unsecured or poorly secured devices.

5.5.2.8 VSAT

This is widely used today for remote monitoring and control of transmission and distribution substations. It is a proven technology for quick implementation but relatively expensive. Major disadvantage is that severe weather will impacts its liability of VSAT. VSAT systems provide high speed, broadband satellite communications for Internet or private network communications. VSAT is ideal for remote monitoring and control such as off- shore wind farms, mining industry, vessels at sea, oil and gas camps or any application that requires a broadband Internet connection at a remote location.

Technically, a VSAT is a two-way satellite ground station with a dish antenna that is smaller than 3.8 meters. The majority of VSAT antennas range from 79cm to 1.2m. Data rates, usually, range from 4 Kbit/s up to 16 Mbit/s. VSATs access satellites in geosynchronous orbit or geostationary orbit to relay data from small remote Earth stations to other stations in mesh topology or master Earth station *hubs* in star topology. VSATs can be used to transmit both narrowband data and broadband data. It also uses portable, phased array antennas or mobile communication infrastructures.

5.6   SECURITY IN WIRELESS COMMUNICATIONS

Wireless communication is a rapidly expanding field of technology for networking, connectivity, communication, and data exchange. There are literally thousands of protocols, standards, and techniques that can be labeled as wireless. These include cell phones, Bluetooth, cordless phones, and wireless networking. As wireless technologies continue to proliferate, the organization's security efforts must go beyond locking down its local network. Security should be an end-to-end solution that addresses all forms, methods, and techniques of communication. While managing network security with filtering devices such as firewalls and proxies is important, one must not overlook the need for endpoint security. Endpoints are the ends of a network communication link. One end is often at a server where a resource resides, and the other end is often a client making a request to use a network resource. Even with secured communication protocols, it is still possible for abuse, misuse, oversight, or malicious action to occur across the network because it originated at an endpoint. All aspects of security from one end to the other, often called *end-to-end security*, must be addressed.

Any unsecured point will be discovered eventually and abused. Endpoint security is the security concept that encourages administrators to install firewalls, malware scanners, and IDS on every host.

5.6.1 ENDPOINT THREAT DETECTION AND RESPONSE (ETDR)

Endpoint security is the concept that each individual device must maintain local security whether it is a network or telecommunications channels. In a modest way it can be referred as the protection of an organization's network when accessed via remote devices such as laptops or other wireless or mobile devices. Sometimes this is expressed as the end device which is responsible for its own security. However, a clearer perspective is that any weakness in a network, whether on the border, on a server, or on a client, presents a risk to all elements within the organization. Traditional security was depended on the network border sentries, such as appliance firewalls, proxies, centralized virus scanners, and even IDS/IPS/IDP solutions, to provide security for all of the interior nodes of a network. This is no longer considered best business practice because of internal and external threats. The security of the network is ultimately decided by the weakest element in the network. Lack of internal security is even more problematic when remote access services, including dial-up, wireless, and VPN, might allow an external entity (authorized or not) to gain access to the private network without having to go through the border security gauntlet. Endpoint security should therefore be viewed as an aspect of the effort to provide sufficient security on each individual host. Every system should have an appropriate combination of a local host firewall, antimalware scanners, authentication, authorization, auditing, spam filters, and IDS/IPS services.

Endpoint Threat Detection and Response (ETDR) mainly focus on the endpoint threats. It includes incident response and collection of tools mainly used for detection and incident response.

5.6.2 TRANSPARENCY

Just as the name implies, transparency is the characteristic of a service, security control, or access mechanism that ensures that it is unseen by users. Transparency is often a desirable feature for security controls. The more transparent a security mechanism is, the less likely a user will be able to circumvent it or even be aware that it exists. With transparency, there is a lack of direct evidence that a feature, service, or restriction exists, and its impact on performance is minimal. In some cases, transparency may need to function more as a configurable feature than as a permanent aspect of operation, such as when an administrator is troubleshooting, evaluating, or tuning a system's configurations. A security boundary can be the division between two secured areas, or it can be the division between a secured area and an unsecured area. But both must be addressed in a security policy.

5.6.3   REDUNDANCY IN SCADA AND SMART GRID

In order to make the industrial SCADA systems most reliable, it is necessary to have redundancy built into the SCADA. Satellite Communications not only provide a feasible means for distributed SCADA communication but also for redundancy for SCADA as a whole. Given the rate at which the cost for satellite communications is dropping and the available bandwidth is increasing, satellite communications infrastructure can be deployed along with conventional broadband technologies which are used in urban areas. Also, given the fact that satellite communications may not be used most of the time in urban areas, except for times when the main infrastructure goes down, the cost for satellite communications infrastructure for Smart Grid may drop further.

Satellite communications is found very useful in providing redundancy in SCADA especially when employed in PSS. One of the advantages is that it does not require much terrestrial equipment. The only equipment required to establish communications are the VSATs and modems. The main advantage of employing the satellite communication is that even during the adverse weather conditions or disasters like floods, systems like cable and Wi-Fi fail because of wires or repeaters being knocked down, satellite communications would still continue to function. This is especially important since utilities might need to know the field conditions of the distributed plant or process during such times, to find faults or problems in the industry and fix the solutions.

VSATs can be run on backup batteries, if the main source of its power goes down and still continue to relay information to the utilities. In this respect satellite communications is the only technology that can efficiently deliver the results during times of disasters and rough weather conditions.

One of the fears in the past is the poor satellite communications connectivity during thick cloud cover or during rains or heavy snow.

SUMMARY

This chapter has focused on various communication aspects of the industrial SCADA with an emphasis on DCS and Smart Grid. It begins with discussing various types of transmission technology in very modest way so that it is very apt and most essential for a power engineer who is engaged in the design and implementation of SCADA and Smart Grid. The chapter then discusses the guided and unguided media used today for communication in such a manner that it is very useful for a practicing communication professional, which includes the various cabling issues as well. Various but most relevant communication technologies, which find space not only in industrial SCADA but also in other smart automation technologies today are discussed comprehensively. Finally the chapter focused on the security issues of the wireless communication technology.

Chapter SIX

SCADA AND DCS PROTOCOLS

6.1 INTRODUCTION

To obtain full functionality in SCADA, especially in DCS it needs appropriate protocols for transmitting data between its components. SCADA, being the heart and brain of ICS and Smart Grid, standardized and interoperable SCADA protocols are very much needed for their efficient and smooth operation. The old SCADA communication protocols such as Modbus RTU, though Profibus and Conitel are widely adopted and used, they are SCADA-vendor specific. Standard protocols which are becoming prevalent today are IEC 61850, IEC 60870-5-101 or 104, and DNP3. These communication protocols are standardized and adopted by almost all major SCADA vendors. Many of these protocols are now considerably modified and contain extensions to operate over TCP/IP as well. However it is a good security engineering practice to avoid connecting SCADA systems to the internet so that the attack surface can be considerably reduced. RTUs and other automatic controller devices were being developed before the advent of industry wide standards for interoperability. This results in multitude of control protocols. This chapter is dedicated to explain various protocols which are relevant to SCADA and DCS environment with an emphasis on cyber-security.

6.2 EVOLUTION OF SCADA COMMUNICATION PROTOCOLS

SCADA protocols evolved out of the need to send and receive data and control information locally and over distances in deterministic time. Deterministic in this context refers to the ability to predict the amount of time required for a transaction to take place when all relevant factors are known and understood. To accomplish communication in deterministic time for applications in ICS and DCS systems, manufactures developed their own protocols and communication bus structure. Profibus of Siemens and Modbus of Modicon which is presently owned by Schneider Electric, are typical examples. Since all the protocols are proprietary, interoperability became a major challenge. This made the control industries and standards organizations to develop open SCADA protocols for control systems which would be nonproprietary and not exclusive to one manufacturer. Further as the internet gained popularity, manufactures started incorporating the protocols and tools which are developed internet such as TCP/IP series of protocols and Internet browsers. In addition, manufactures and open standards organizations modified the highly popular and efficient Ethernet LAN technology in implementing data acquisition and control networks.

De jure standards are developed by national and International standards development organizations such as ANSI, IEEE, NIST, IEC, etc. Many de facto industrial standards are made de jure, after appropriate evaluation. Modbus is one of the typical such SCADA communication protocol extensively deployed today. The following sections discuss a few of the popular communication protocols used in DCS scenario. ICCP (IEC 60870-6) is the international standard for one control center to communicate with another control center of Power System SCADA. For communication from master stations to substations DNP3 is used in North America and IEC T-101 serial and T-104 (TCP/IP) are used in Europe. For communication between field equipment, IEC 61850, DNP3 (IEEE1815) and Modbus are developed.

6.3   SCADA COMMUNICATION PROTOCOLS

As explained above, the use of international open protocol standards are now recognized throughout the industry especially in electric utility as a key to successful integration of the various parts of the electric utility enterprise. Benefits of open systems include longer expected system life, investment protection, upgradeability and expandability, and readily available third-party components. The following sections elaborate some of the important communication protocols which are presently used by many industries and power utilities.

6.3.1 DISTRIBUTED NETWORK PROTOCOL (DNP) 3.0

DNP3 is extensively used in many industries in automation like electricity, oil and gas, and water. It is popular and extensively used in the United States of America, Canada, South America, Australia, and parts of Asia and Africa. It is a set of an open SCADA protocol that is used for serial or IP communication between control devices and initially developed by Westronics in Calgary, Alberta, Canada. DNP3 is mainly used between components in process automation systems especially in SCADA systems employed in electric and water companies, but usage in other industries is not common. It has larger data frames and can carry larger RTU messages, and is very much useful for communications between various types of data logging and control devices such as Remote Terminal Units (RTUs), Data Concentrators and Intelligent Electronic Devices (IEDs). It is specifically designed to achieve reliability and efficiency when used in real-time data transfer. Another feature is that it supports time-stamped data communications between a master station and RTUs or IEDs or Phasor Measurement Units (PMU).

DNP3 faces the similar cyber security problems as IEC 60870-5-104[T-104] and is explained in subsequent sections. Scrutiny made by security engineers for integrity in DNP3 revealed that it lacks authentication and encryption. The DNP3 function codes and data types are well known, hence it can be easy to manipulate and compromise a DNP3 communication session.

6.3.1.1 Protocol Architecture of DNP3

DNP3 is based on the Enhanced Performance Architecture (EPA) and uses the frame format FT3 specified by IEC 60870-5. The lower layers of physical and data link defining the communication between devices are similar to IEC 60870-5-101 and the higher levels of data units and functionality are different. DNP3 uses cyclic redundancy check for error detection.

The DNP3 protocol structure uses the basic three layer EPA model with some added functionality. It adds an additional layer named the pseudo-transport layer. The pseudo-transport layer is a combination of network and transport layer of the Open Systems Interconnection (OSI) model and also includes some functions of the data link layer. Network function is concerned with the routing and data flow over the network from sender to receiver. Transport function includes proper delivery of the message from sender to receiver, message sequencing, and corresponding error correction. This function of a transport layer is limited when compared to the OSI layer, and hence the name pseudo-transport layer, as shown in Figure 6.1.

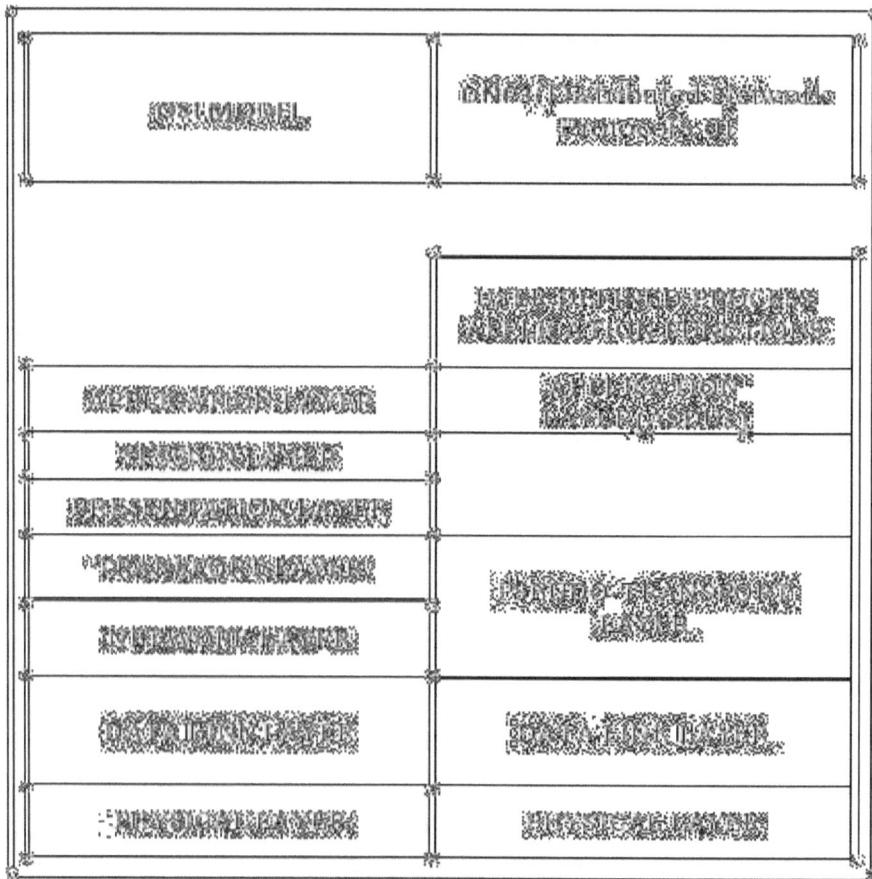

Figure 6.1   DNP3 and OSI model layers

6.3.2 MODBUS

Modbus is a request-response serial communications protocol implemented using a master-slave relationship. In a master-slave relationship, communication always occurs in pairs. Which means one device (the slave) must initiate a request and then wait for a response, and the initiating device (the master) is responsible for initiating every interaction is shown in Figure 6.2. Typically, the master is a Human Machine Interface (HMI) or SCADA server and the slave is a sensor, RTU, PLC, or Programmable Automation Controller (PAC). The content of these requests and responses, and the network layers, across which these messages are sent, are defined by the different layers of the protocol.

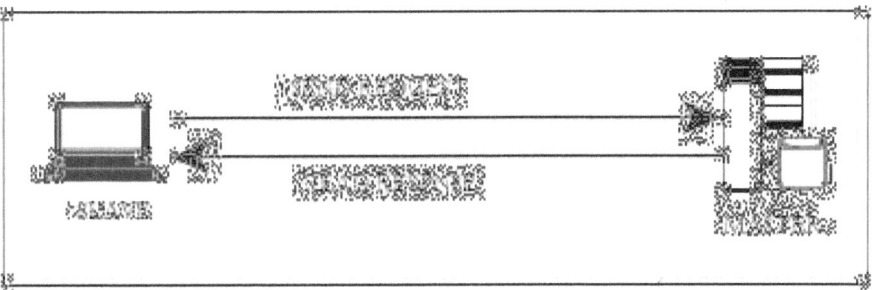

Figure 6.2   A Master-Slave Networking Relationship

Modbus is originally published by Modicon which is presently owned by Schneider Electric for use in their PLCs. Modbus being simple and robust, become a *de facto* standard communication protocol. It is now commonly available and used for connecting industrial electronic devices in ICS. The main reasons for the use of Modbus in the industrial environment are:

1. developed with industrial applications in mind,

2. openly published and royalty-free,

3. easy to deploy and maintain, and

4. moves raw bits or words without placing many restrictions on vendors.

Modbus enables communication among many devices connected to the same network, for example, a system that measures temperature and humidity and communicates the results to a computer. Modbus is often used to connect a supervisory computer with RTUs in SCADA systems. Many of the data types are named from its use in driving relays: a single-bit physical output is called a coil, and a single-bit physical input is called a discrete input or a contact.

The development and update of Modbus protocols has been managed by the Modbus organization, since Schneider Electric transferred rights to that organization. The Modbus organization is an association of users and suppliers of Modbus-compliant devices that seek to drive the adoption and evolution of Modbus. The different versions of the Modbus protocol which are presently used are:

1. Modbus RTU

2. Modbus ASCII

3. Modbus TCP/IP or Modbus TCP,

4. Modbus over UDP,

5. Modbus over TCP/IP or Modbus over TCP or Modbus RTU/IP,

6. Modbus over UDP Modbus Plus (Modbus+, MB+ or MBP),

7. Pemex Modbus, and

8. Enron Modbus.

Among these Modbus RTU, Modbus ASCII, and Modbus TCP/IP or Modbus TCP are the most popular versions. A brief description and the frame format of these versions are summarized below.

1. *Modbus RTU:* This is the most common implementation available for Modbus and is used in serial communication by making use of a compact, binary representation of the data for protocol communication. The RTU format follows the commands/data with a cyclic redundancy check (CRC) checksum as an error check mechanism to ensure the reliability of data. A Modbus RTU message must be transmitted continuously without inter-character hesitations. Modbus messages are framed (separated) by idle (silent) periods.

    A Modbus frame is composed of an Application Data Unit (ADU), which encloses a Protocol Data Unit (PDU):

    1. ADU = Address + PDU + Error check,

    2. PDU = Function code + Data.

    Modbus RTU frame format (primarily used on 8-bit asynchronous lines like EIA-485) and can be represented as shown below:

Start -28 bits	Address 8 bits	Function 8 bits	Data n x 8 bits	CRC 16 bits	End 28 bits

2. *Modbus ASCII*:   This is used in serial communication and makes use of ASCII characters for protocol communication. The ASCII format uses a longitudinal redundancy checksum. Modbus ASCII frame format (primarily used on 7- or 8-bit asynchronous serial lines) is given below.

Start -8 bits	Address 16 bits	Function 16 bits	Data n x 2 bits	LRC 16 bits	End 16 bits

3. *Modbus TCP/IP or Modbus TCP:*   This is a Modbus variant used for communications over TCP/IP networks, connecting over port 502. It does not require a checksum calculation, as lower layers already provide checksum protection. Modbus TCP frame format is given below.

Transaction identifier 16 bits	Protocol identifier 16 bits	Length field 16 bits	Unit identifier 8 bits	Function code 8 bits	Data bytes n bits

One of the other Modbus variants gaining popularity is the Modbus plus. It is a proprietary to Schneider Electric and unlike the other variants, it is a high speed peer to peer network protocol based on token passing communication. It requires a dedicated co-processor to handle fast HDLC-like token rotation. It uses twisted pair at 1 Mbit/s and includes transformer isolation at each node, which makes it transition/edge-triggered instead of voltage/level-triggered. Special hardware is required to connect Modbus Plus to a computer, typically a card made for the ISA, PCI or PCMCIA bus.

In Modbus, data transactions are traditionally stateless, making them highly resistant to disruption from noise. It requires minimal recovery information at either end. Programming operations, on the other hand, expect a connection-oriented approach. This was achieved on the simpler variants by an exclusive login token, and on the Modbus Plus variant by explicit Program Path capabilities which maintained a duplex association until explicitly broken down. The main reason why the connection-oriented TCP/IP protocol is used is to keep control of an individual transaction by enclosing it in a connection, which can be identified, supervised, and cancelled without requiring specific action on the part of the client and server applications. This gives the mechanism of wide tolerance to network performance changes, and allows security features such as firewalls and proxies to be easily added. The other versions of Modbus are briefly explained below.

1. *Modbus over UDP:* Modbus over UDP on IP networks removes the overheads required for TCP.

2. *Enron Modbus:* This is another extension of standard Modbus developed by Enron Corporation with support for 32-bit integer and floating-point variables and historical and flow data. Data types are mapped using standard addresses.

3. *Pemex Modbus:* This is an extension of standard Modbus with support for historical and flow data. It was designed for the Pemex oil and gas company for use in process control but never gained widespread adoption.

Data model and function calls are identical for the Modbus RTU, Modbus ASCII, Modbus TCP/IP, and Modbus over UDP variants, but the encapsulation is different. However the variants are not interoperable.

6.3.2.1   Modbus Limitations

1. Modbus protocol does not provide any security against unauthorized commands.

2. Modbus was designed and developed initially for the communication with the PLC. Hence the number of data types was limited to those which understood by PLCs at that time.

3. No standard way exists for a node to find the description of a data object, such as determining whether a register value represents a temperature between T1 and T2 degrees.

4. Since Modbus is a master-slave protocol, there is no way for a field device to *report by exception.* As a result the master node must routinely poll each field device and look for changes in the data. This consumes considerable bandwidth and network time in applications.

5. Modbus is restricted to addressing 254 devices on one data link, which limits the number of field devices that may be connected to a master station.

6. Modbus transmissions must be contiguous, which limits the types of remote communications devices to those that can buffer data to avoid gaps in the transmission.

6.3.2.2   Attacks on the DNP3 and Modbus

As the original focus while developing these was not security rather efficiency, Modbus and DNP3 virtually have no security measures incorporated. The lack of authentication in Modbus means that remote terminals accept commands from any machine that appears to be a master. The lack of integrity checking or encryption allows messages to be intercepted, changed and forwarded. In fact it is susceptible to Man In The Middle attack (MITM) which is most perilous in ICS and DCS which control critical infrastructure. For DNP3, message from an outstation can easily be spoofed, making it appear to be unavailable to the master. It is also common that passwords may be sent across the network in clear text. Though the DNPSecure has been developed, deployment is difficult as most of the systems using DNP3 are in the Critical Infrastructure, where downtime is very crucial and almost not feasible.

Modbus protocol is highly vulnerable to attacks like response and measurement injection, and command injection. A scrutiny on Modbus protocol with various levels of injection attacks ranging from naïve injection to complex injections targeting specific fields and values based on domain knowledge, reveals the possible consequences, such as sporadic sensor measurements, altered system control schemes and altered actuator states. This can result in partial communication disruption to complete shutdown of the device.

6.3.3   PROFIBUS

Profibus is an open standard, serial, smart field-bus technology widely used in time-critical control and data acquisition systems. It can provide data transmission rates of 31 Kbps, 1Mbps, and 2,5 Mbps in the physical layer.

It provides determinism for real-time control applications and supports multimaster communication networks.

Profibus uses the bus topology. In this topology, a central line, or bus, is wired throughout the system. Devices are attached to this central bus. One bus eliminates the need for a full-length line going from the central controller to each individual device. In the past, each Profibus device had to connect directly to the central bus. Technological advancements, however, have made it possible for a new *two-wire* system. In this topology, the Profibus central bus can connect to a ProfiNet Ethernet system. In this way, multiple Profibus buses can connect to each other.

Profibus devices which are connected to a central line can communicate information in an efficient manner which can go beyond the automation messages. Profibus devices can also participate in self-diagnosis and connection diagnosis. At the most basic level, Profibus benefits from superior design of its OSI layers and basic topology. There are three versions of Profibus, which are summarized below.

6.3.3.1   Profibus Process Automation (PA)

Profibus PA is a protocol designed for Process Automation. In fact, Profibus PA is a type of Profibus DP (Decentralized Peripherals) application profile. It connects data acquisition and control devices on a common serial bus and supports reliable, intrinsically safe implementations. It also provides power to field devices through the bus. Profibus PA standardizes the process of transmitting measured data. It does hold a very important unique characteristic. Profibus PA was designed specifically for use in hazardous environments. Profibus uses the basic functions and extensions available in Profibus DP.

In most environments, Profibus PA operates over RS485 twisted pair media. This media, along with the PA application profile supports power over the bus. In explosive environments, though, that power can lead to sparks that induce explosions. To handle this, Profibus PA can be used with Manchester Bus Powered technology (MBP). The MBP media was designed specifically to be used in Profibus PA. It permits transmission of both data and power. The stepping down of power, reduces, or nearly eliminates, the possibility of explosion. Buses using MBP can reach 1900 meters and can support branches.

6.3.3.2   Profibus Factory Automation (Decentralized Peripherals-DP)

The second type of Profibus is more universal which is referred as Profibus DP, for Decentralized Periphery, this new protocol is much simpler and faster. Profibus DP is most popular and has an overwhelming majority of Profibus application today. It uses different physical layer standards than those employed by Profibus PA. Application profiles allow users to combine their requirements for a specific solution. Profibus DP itself has, three separate versions. Each version, from DP-V0 to DP-V1 and DP-V2, provides newer, more complicated features such as diagnostics, alarm messaging, and parameterization.

6.3.3.3   Profibus Fieldbus Message Specification (FMS)

The initial version of Profibus was Profibus FMS, Fieldbus Message Specification. Profibus FMS was designed to communicate between Programmable Controllers and PCs, sending complex information between them. Unfortunately, being the initial effort of Profibus designers, the FMS technology was not as flexible as needed. This protocol was not appropriate for less complex messages or communication on a wider and complicated network. Though it offers a large number of functions and is, generally, more complicated to implement than Profibus PA or Profibus DP. New types of Profibus would satisfy those needs. Profibus FMS is still in use today, though the vast majority of users find newer solutions to be more appropriate.

A comparison of the three Profibus versions is given in Figure 6.3 below.

Profibus PA	Profibus DP	Profibus FMS
Intrinsic safety, Reliable, Bus Powered, Process Applications	High speed, Decentralized Applications	General Automation, Large number of Application

Figure 6.3   Profibus Versions

6.3.3.4 Communication Architecture of Profibus

Figure 6.4 illustrate the communication architecture of the Profibus versions and shows their relationships in the OSI seven layer model.

OSI MODEL	Profibus PA	Profibus DP	Profibus FMS
Application Layer	application layer not used	application layer not used	application layer field bus message specification
Session Layer			
Presentation Layer			
Transport Layer			
Network Layer			
Data Link Layer	data link layer IEC interface	data link layer fieldbus data link	data link layer fieldbus data link
Physical Layer	physical layer IEC 61158-2	physical layer eia-485, fiber optics, radio waves	physical layer EIA-485, fiber optics, radio waves

Figure 6.4   Profibus PA, DP, and FMS layered protocols.

Profibus systems can have three types of physical media. The first is a standard twisted-pair wiring system, in this case RS485. Two more advanced systems are also available. Profibus systems can now operate using fiber-optic transmission in cases where that is more appropriate. A safety-enhanced system called Manchester Bus Power, or MBP, is also available in situations where the chemical environment is prone to explosion. The physical layers can also use either the EIA-485 standard or the IEC 61158-2 standard. If desired, all three Profibus versions can use the same bus line if they employ EIA-485 in the physical layer. However, if the application requires intrinsically safe circuitry, IEC 61158-2 operates at 31.25Kbps.

6.3.4 IEC 60870-5-101/103/104

IEC 60870 was introduced by the IEC Technical Committee 57 and widely used in ICS and DCS communication. It is mainly popular in Europe and China and suitable for controlling electric power transmission grids and other geographically widespread ICS. This is an open protocol originally written for serial communication and was released in 1995. The IEC 60870-5-104 standard, released in 2000, present a combination of the application layer of IEC 60870-5-101 and the transport functions provided by TCP/IP. Presently this is widely used in SCADA and DCS communication protocol especially for monitoring and controlling of remote field locations of DCS especially in electrical grids. The structure of IEC 60870-5 standard is hierarchical and has six parts, each having different sections, and has four companion standards. Main part of the standard defines the fields of application, whereas the companion standards elaborate the information regarding the application field by specific details. These companion standards may be referred as T-101, T-102, T-103, and T-104, where T stands for Telecontrol. The five documents specify the base IEC 60870-5 Tele-control equipment and systems. The six sections of Part 5 of the IEC 60870-5 are described below.

1. IEC 60870-5-1 Transmission Frame Formats
2. IEC 60870-5-2 Data Link Transmission Services
3. IEC 60870-5-3 General Structure of Application Data
4. IEC 60870-5-4 Definition and Coding of Information Elements
5. IEC 60870-5-5 Basic Application Functions
6. IEC 60870-5-6 Guidelines for conformance testing for the IEC 60870-5 companion standards
7. IEC TS 60870-5-7 Security extensions to IEC 60870-5-101 and IEC 60870-5-104 protocols (applying IEC 62351)

6.3.5 IEC 60870-5-101 [T-101]

An Enhanced Performance Architecture (EPA) based architecture released by IEC in the beginning of 90s, which found extensive acceptance in power system SCADA. It is a master slave communication for multi-drop or bus topology. It polls data by cyclic polling technique using the data link layer and use parity check as well as checksum error detection techniques. The data error is decreased by these checks in IEC 60870 protocols. This standard is widely used in power system communications for telecontrol, teleprotection, and associated telecommunications. The application function of T-101 includes station initiation, parameter loading, data acquisition, cyclic data transmission, clock synchronization, etc. for a remote substation. This is completely compatible with IEC 60870-5-1 to IEC 60870-5-5 standards and uses standard asynchronous serial telecontrol channel interface between master and slave. The standard is suitable for multiple configurations like point-to-point, star, multidropped etc. The specific features of IEC 608705-101 are described below.

1. Both master initiated and master/slave initiated modes of data transfer are available,

2. It functions based on Link address and Application Service Data Unit (ASDU) bit,

3. ASDU addresses are provided for classifying the end station,

4. Data can be classified into different information objects and each information object is provided with a specific address,

5. Capability to assigning priority to the data before transmitting the data,

6. Option of classifying the data into 16 different groups to get the data according to the group by issuing specific group interrogation commands from the master,

7. Capability for time synchronization, and

8. Various schemes for data transfer are available which is very much useful in IED's to transfer the SoE and disturbance files for fault analysis.

6.3.6   IEC 60870-5-103 [T-103]

This is a master-slave standard protocol based on the EPA architecture and is mainly used to couple the central unit to several protection devices and is primarily used in the energy sector. This standard is especially designed for the communication with protection devices and therefore difficult to adapt it to other applications. It defines a companion standard that enables interoperability between protection equipment and devices of a control system in a substation. It handles protection functions as status indications of circuit breakers, types of fault, trip signals, auto-reclosure, relay pickup, etc. The device complying with this standard can send the information using two methods for data transfer - either using the explicitly specified ASDU or using generic services for transmission of all the possible information. The standard supports some specific protection functions and provides the vendor a facility to incorporate its own protective functions on private data ranges. IEC 60870-5-103 is mostly used for comparatively slow transmission media with RS232 and RS485 interfaces. Connection via optical fiber is also covered by the standard. The transmission speed in general is specified with a maximum of 19200 Baud.

6.3.7   IEC 60870-5-104 [T-104]

IEC 60870-5-104 [T-104] protocol is a standard for telecontrol equipment and systems with coded bit serial data transmission in TCP/IP based networks for monitoring and controlling geographically widespread processes. It is an extension of IEC 101 protocol suitable for networked circumstances with the changes in transport, network, link and physical layer services to suit the complete network access. The standard uses an open TCP/IP interface to network to have connectivity to the Local Area Network (LAN) and routers with different facility used to connect to the Wide Area Network (WAN). Within TCP/IP various network types can be utilized including X.25, Frame Relay, ATM, ISDN, Ethernet and serial point to point (X.21), Application layer of IEC 104 is preserved same as that of IEC 101 with some of the data types and fascilities which are not used. There are two separate link layers defined in the standard, which is suitable for data transfer over Ethernet and serial line Point-to-Point Protocol (PPP). The control field data of IEC104 contains various types of mechanisms for effective handling of network data synchronization.

Unfortunately, by design itself the security of IEC 104 is problematic, same as many of the other contemporary SCADA protocols developed. IEC Technical Committee (TC) 57 have published a security standard IEC 62351, which implements end-to-end encryption which would prevent such attacks as replay, man-in-the-middle and packet injection. However due to the increase in complexity and cost, vendors are reluctant to include this security features on their ICS networks.

6.3.7.1   Protocol Architecture of IEC 60870-5

Protocol Architecture of IEC 60870 is based on Enhanced Performance Architecture (EPA) model. The EPA model has three layers viz. Physical, Data Link and Application layers. An user layer is added to the top of the EPA model to provide the interoperability between equipements in a telecontrol system. This four layer model is used for T-101, and T-103 companion standards. For companion standard T-104, which is the network adaptation, some additional layers are included from the OSI model. These are network and transport layers that are essential for the networked architecture. This networked architecture is useful for the transportation of data and messages over the network. Thus a non networked version is used for T-101, T-103, and the networked version for T-104, as shown in Figure 6.5. It may be noted that the lower four layers of T-104 are now the TCP/IP suite for networking applications.

OSI model	T-101, T-103 (EPA with user defined process functions)	T-104- EPA with network and transport with user defined process functions
	User defined process application functions	User defined process application functions
Application layer	Application layer (ASDUs)	Application layer (ASDUs)
Session layer		
Presentation layer		
Transport layer		TCP/IP transport and network protocol suite

Network layer		TCP/IP transport and network protocol suite
Data link layer	Data link transmission procedures and frame formats	
Physical layer	Physical interface specification	

Figure 6.5 Communication layers of IEC 60870-5

6.3.7.2 Attacks on IEC 60870-5

SCADA StrangeLove demonstrates how to detect IEC 60870 device on the network. They also released python scripts which can identify and return the common address of an IEC 60870 device. The common address is an address used for all data contained within the IEC 60870-5 packet, used to identify the physical device. Using these existing scripts it would be possible to scan a network for specific IEC 60870 hosts. Once reconnaissance became successful, the scripts could be used to detect possible targets for a man in the middle attack, after confirming the end device is using the IEC 60870-5 protocol and obtaining its address. Improperly terminated VPN running on IEC 60870-5 protocols is highly susceptible to DoS attacks, hence end node problem has to be sorted out with utmost care.

Note: SCADA StrangeLove is an independent group of cyber security engineers founded in 2012, focused on security assessment of ICS and SCADA.

6.3.8 IEC 61850

Multiple protocols exist for electrical substation automation, which include many proprietary protocols with custom communication links. Interoperation of devices from different vendors is a need of the hour in power system automation, as it would be indeed an advantage to the designers of substation automation. This brought a need for a robust interoperable standard to serve the power system SCADA. This requirement has been attended by a group of about 60 members from different countries worked in three IEC working groups from 1995 and created IEC 61850 which accomplished the following objectives.

1. A single protocol for complete substation automation considering all different data transfer requirement,

2. Definition of basic services required to data transfer,

3. Provide high inter-operability between systems from different vendors,

4. A common method/format for storing complete data, and

5. Define complete testing required for the equipment which conforms to the standard.

IEC 61850 is also a layered architecture standard that separates the functionality required for electric utility applications from the lower level networking tasks. The layered architecture illustrating the separation of functions is shown in Figure 6.6.

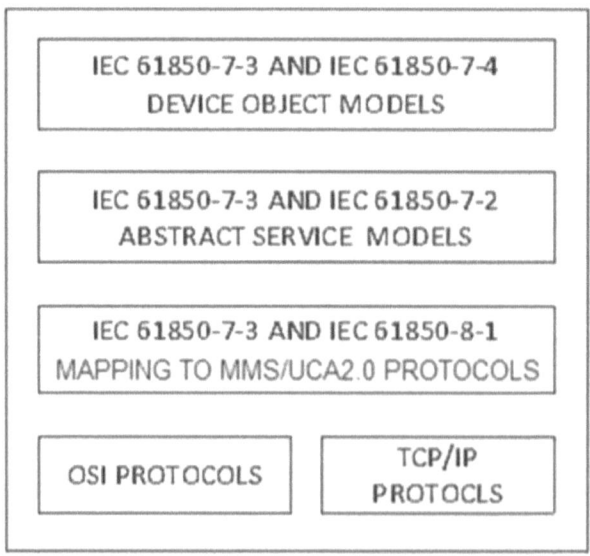

Figure 6.6   IEC 61850 Layered Architecture

IEC 61850 is an object oriented substation automation standard which defines how to describe the devices in an electrical substation and how to exchange the information about these devices. The information model of

IEC 61850 is based on two main levels of modelling. The breakdown of physical device into logical device, and the breakdown of logical device into logical nodes, data objects and data attribute. The approach of IEC 61850 is to decompose the application function into the smallest entities which are used to exchange information.

While implementing the PSS, the data exchange between the process level devices and substation is necessary. In order to achieve this all participating devices must be compatible with this protocol. As the IEC 61850 standard has been developed with a clear vision to fully incorporate the IED interoperability, irrespective that the IEDs belong to different vendors. In fact the IEC 61850 compliance of devices takes the substation automation to a next level. IEC 61850 offers three types of communication models.

1. Client/server type communication services model,

2. A publisher subscriber model, and

3. Sample Value model for multicast measurement values.

Generic Object Oriented Substation Events: Generic Object Oriented Substation Events (GOOSE) is a controlled model mechanism in which any format of data (Status Value) is grouped into a data set and transmitted within a time period of 4 milliseconds. The message structure of GOOSE supports the exchange of a wide range of possible common data organized by a dataset. Being the GOOSE message a multicast, it is received by all the IEDs which have been connected and configured to subscribe it. GOOSE message usually carries the information of a status changed and time of the last status change. In other way, GOOSE time stamped events from IEDs are appropriately communicated. The message frame consists of the destination/source Media Access Control (MAC), addresses (DEST/ SRC), tag protocol identifies, length, two reserved fields and application protocol data unit.

6.3.8.1 Comparison of DNP3 vs. IEC- 61850 GOOSE

IEC 61850 is not just a communication protocol as it not only defines how the data is transmitted and received but also describes how data is executed and stored. This makes data model of IEC 61850 very different to the OSI reference model. This difference in the data model requires IEC61850 to work over a real set of communication protocols such as DNP3 or IEC T[101/3/4], requiring the data model of IEC 61850 to be mapped on to one of the above mentioned protocols. IEC 61850 is more advantageous to use only within a substation and its specification states that connection to the remote control centers such as remote MCC is beyond its scope. The use of IEC 61850 would be most appropriate if it is used in a substation environment where a number of IEDs interact with the SCADA master. In fact it is referred as substation protocol.

6.3.8.2   Attacks on IEC 61850 Protocol

Although IEC 61850 based automated substations can provide various advantages over traditional substations, the power supplier companies are very much cautious about its implementation due to the security concerns. Security Engineers have identified a number of security vulnerabilities and flaws in the IEC 61850 protocol such as the lack of encryption used in the GOOSE messages, lack of Intrusion Detection System (IDS) implementation in IEC 61850 networks, and no firewall implementation inside IEC 61850 substation network. If these vulnerabilities are deviously exploited, a number of security attacks can be launched on IEC 61850 substation network.

One of the vulnerabilities within the GOOSE communication of IEC 61850 that can be exploited is by sending GOOSE frames containing higher status numbers. It prevents genuine GOOSE frames from being processed. This effectively causes a hijacking of the communication. This attack could be used to implement a Denial of Service (DoS) attack. This weakness in GOOSE can be exploited to insert spoofed messages with incorrect data between each valid message. This can be used to demonstrate using Scapy, which is a Python program that enables the user to sniff, dissect, forge, and send network packets. This attack is possible due to unencrypted and unauthenticated nature of the GOOSE message.

6.3.9   ICCP TASE .2 (IEC 60870-6)

The Inter Control Center Communications Protocol (ICCP or IEC 60870-6/TASE.2) is the protocol used by various power utilities throughout the world to provide data exchange over Wide Area Networks (WANs) between utility control centers, utilities, power pools, regional control centers, and Non-Utility Generators. In fact today ICCP is the international standard adopted by IEC as ICCP Telecontrol Application Service Element 2 (TASE.2) and is an essential protocol in PSS. Usually a typical national power grid includes a hierarchy of control centers to manage the generation, transmission, and distribution of power throughout the grid. The grid is controlled by one or more hierarchical control centers, which are responsible for scheduling of power generation to meet customer demand, and for managing major network outages and faults, are briefly described below.

6.3.9.1   ICCP Functionalities

The basic functionalities of the ICCP is to establish an appropriate link between the following control centers by managing and configuring it for proper information exchange.

1. Generation control centers, responsible for managing the operation of generating plants such as coal-fired, natural gas, nuclear, solar, wind, etc. and for adjusting the power generated according to the requirements of the system control center,

2. Transmission control centers, responsible for the transmission of power from generating stations to network distributors, and

3. Distribution control centers, responsible for the distribution of power from the transmission networks to individual consumers.

6.3.9.2   Protocol Architecture of ICCP TASE 2

At present, the ICCP TASE 2 protocol is the internationally recognized standard for communications between electrical utility control centers. ICCP uses the Manufacturing Messaging Specification (MMS) for the messaging service needed by ICCP. It is based on client/server principles. Consequently data transfers initiated with a request from one control center (client) to another control center (server). Control centers may be both clients and servers. ICCP TASE 2 operates at the application layer in the OSI model. Any physical interfaces transport and network services that fit this model are supported. However TCP/IP over Ethernet seems to be the most common. ICCP may operate over a single point-to-point link between two control centers. The logical connections or *associations* between control centers are completely general. A client can establish association with more than one server and a client can also establish more than one association with the same server. Multiple associations with same server can be established at different levels of quality of service so that high priority real-time data is not delayed by lower priority or non-real-time data transfers.

6.3.9.3   Implementation Issues and Interoperability

ICCP is a standard real-time data exchange protocol. It provides numerous features for the delivery of data, monitoring of values, program control and device control. All the protocol specifics needed to ensure interoperability between different vendor's ICCP products have been included in the specifications. The ICCP specifications, however, do not attempt to specify other areas that will need to be implemented in an ICCP software product but that do not affect interoperability. These areas are referred to as local implementation issues in the specification. Some of the local implementation issues in the specification are listed below:

1. The API through which local applications interface to ICCP to send or receive data,

2. A user interface to ICCP for user management of ICCP data links,

3. Management functions for controlling and monitoring ICCP data links,

4. Failover schemes where redundant ICCP servers are required to meet stringent availability requirements, such as those typically experienced in an EMS/SCADA system environment, and

5. How data, programs or devices will be controlled or managed in the local SCADA/EMS to respond to requests received via an ICCP data link.

The wide acceptance of ICCP by the utility industry has resulted in several ICCP products are available on the market. Extensive interoperability testing between products of some of the major vendors has been a feature of ICCP protocol development. An ICCP purchaser must define functionality required in terms of conformance blocks and the objects within those blocks. Application profiles for the ICCP client and server conformances must match if the link is to operate successfully. Interoperability among ICCP of different vendors for the power grid is crucial for achieving the benefits of standardization such as application evolution, open architecture and scalability, plug and play capability of components and services, reliability and service orientation.

6.3.9.4   ICCP-Product Differentiation

ICCP is a real-time data exchange protocol providing features for data transfer, monitoring and control. For a complete ICCP link there need to be facilities to manage and configure the link and monitor its performance. The ICCP standard does not specify any interface or requirements for these features that are necessary but nevertheless do not affect interoperability. Similarly failover and redundancy schemes and the way the SCADA responds to ICCP requests is not a protocol issue, so is not specified. These non-protocol specific features are referred in the standard as local implementation issues. ICCP implementers are free to handle these issues as per their requirements. Local implementation means that developers have to differentiate their product in the market with added features and values. The money spent for a product with appropriately developed maintenance and diagnostic tools will be valued, during its life expectancy, only if the ICCP connection grow and adapt the changes.

6.3.9.5   ICCP- Product Configurations

Commercial ICCP products are generally available for one of three configurations:

1. as a native protocol embedded in the SCADA host,
2. as a networked server, and
3. as a gateway processor.

As an embedded protocol the ICCP management tools and interfaces are all part of the complete suite of tools for the SCADA. This configuration offers maximum performance because of the direct access to the SCADA database without requiring any intervening buffering. This approach may not be available as an addition to a legacy system. The ICCP application may be restricted to accessing only the SCADA environment in which it is embedded. A networked server making use of industry standard communications networking to the SCADA host may provide, performance approaching that of an embedded ICCP application. On the application interface side the ICCP is not restricted to the SCADA environment but is open to other systems such as a separate data historian or other databases. Security may be easier to manage with the ICCP server segregated from the operational real-time systems. The gateway processor approach is similar to the networked server except it is intended for legacy systems with minimal communications networking capability and so has the lowest performance. In the most minimal situation the ICCP gateway may communicate with the SCADA host via a serial port in a similar manner to the SCADA RTUs. Certain other Standards which are deployed in smart grid and under development are briefly described below.

6.4   OTHER RELEVANT STANDARDS

Some of the protocols which are relevant and important especially in Power System SCADA (PSS) and DCS are briefly introduced below.

6.4.1   IEEE C37.118.1 SYNCHROPHASOR STANDARD

This standard defines synchrophasor measurements. It also defines a data communication protocol, including message formats for communicating the data in real-time. Mainly intend to take measurements at substations, real-time data sent to control center and collect and align data, for sending on to applications or higher level processing.

A little more elaborated, this standard defines synchrophasors, frequency, and Rate of Change of Frequency (ROCOF) measurement under all operating conditions. It specifies methods for evaluating these measurements and requirements for compliance with the standard under both steady-state and dynamic conditions. Time tag and synchronization requirements are included. Performance requirements are confirmed with a reference model, provided in detail. This document defines a phasor measurement unit (PMU), which can be a stand-alone physical unit or a functional unit within another physical unit. This standard does not specify hardware, software, or a method for computing phasors, frequency, or ROCOF.

6.4.2   IEC 61968 STANDARD

IEC 61968 is a series of standards that will define standards for information exchanges between electrical distribution systems. These standards are being developed by Working Group 14 (WG 14) Technical Committee 57 (TC 57) of the IEC. IEC 61968 is intended to support the inter-application integration of a utility enterprise that needs to collect data from different applications that are legacy or new and each has different interfaces and run-time environments. IEC 61968 defines interfaces for all the major elements of interface architecture for Distribution Management Systems (DMS) and is intended to be implemented with middleware services that broker messages among applications.

6.4.3   IEC 61970 STANDARD

The IEC 61970 series of standards deals with the application program interfaces for energy management systems (EMS). The series provides a set of guidelines and standards to facilitate the following.

1. The integration of applications developed by different suppliers in the control center environment;

2. The exchange of information to systems external to the control center environment, including transmission, distribution and generation systems external to the control center that need to exchange real-time data with the control center;

3. The provision of suitable interfaces for data exchange across legacy and new systems.

6.4.4   IEC 62325 STANDARD

IEC 62325 is a set of standards related to deregulated energy market communications, based on the Common Information Model (CIM). IEC 62325 is a part of the IEC Technical Committee 57 (TC57) reference architecture for electric power systems, and is the responsibility of Working Group 16 (WG16) which works on Standards related to energy market communications.

IEC 62325-301 specifies the common information model for energy market communications. The Common Information Model (CIM) is an abstract model that represents all the major objects in an electric utility enterprise typically involved in utility operations and electricity market management. By providing a standard way of representing power system resources as object classes and attributes, along with their relationships, the CIM facilitates the integration of market management system (MMS) applications developed independently by different vendors, between entire MMS systems developed independently, or between an MMS system and other systems concerned with different aspects of market management, such as capacity allocation, day-ahead management, balancing, settlement, etc.

Note: IEC Technical Committee 57 (TC 57) is one of the technical committees of the IEC. TC 57 is responsible for development of standards for information exchange for PSS, distribution automation and teleprotection.

6.4.5   IEC 61508 STANDARD

IEC 61508 is an international standard published by the IEC and is titled Functional Safety of Electrical/Electronic/Programmable Electronic Safety-related Systems. This protocol is intended to be a basic functional safety standard applicable to all kinds of industry. It defines functional safety as part of the overall safety relating to the Equipment Under Control (EUC) and the EUC system which depends on the correct functioning of the E/E/PE safety-related systems, other technology safety-related systems and external risk reduction facilities. The standard covers the complete safety life cycle, and may need interpretation to develop sector specific standards. It has its origins in the process control industry. The safety life cycle of IEC 61508 has 16 phases which can be categorized into three groups as follows.

1.   Phases 1-5 address analysis,

2.   Phases 6-13 address realization, and

3.   Phases 14-16 address operation.

All phases are concerned with the safety function of the system.

IEC 61508:2010 sets out the requirements for ensuring that systems are designed, implemented, operated and maintained to provide the required Safety Integrity Level (SIL). Four SILs are defined according to the risks involved in the system application, with SIL-4 is meant to protect against the highest risks. The standard specifies a process that can be followed by all links in the ICS so that information about the system can be communicated using common terminology and system parameters. The standard is in eight parts which are described below.

1. IEC 61508-0, Functional safety and IEC 61508

2. IEC 61508-1, General requirements

3. IEC 61508-2, Requirements for E/E/PE safety-related systems

4. IEC 61508-3, Software requirements

5. IEC 61508-4, Definitions and abbreviations

6. IEC 61508-5, Examples and methods for the determination of safety integrity levels

7. IEC 61508-6, Guidelines on the application of IEC 61508-2 and IEC 61508-3

8. IEC 61508-7, Overview of techniques and measures

IEC 61508 has been adopted in the UK as BS EN 61508, with the EN indicating adoption also by the European electrotechnical standardization organization CENELEC. Other standards are being produced for the application of the 61508 approach to particular sectors. Sector specific standards related to IEC 61508 include:

1. IEC 61511 Process industries

2. IEC 61513 Nuclear power plants

3. IEC 62061 Machinery sector

4. IEC 61800-5-2 Power drive systems.

6.4.6   IEC 62351 SECURITY STANDARD

IEC 62351 is an industry standard developed for improving security in automation systems especially in the PSS domain. It comprises provisions to ensure the integrity, authenticity and confidentiality for different protocols used in PSS.

IEC 62351 is developed by WG15 (which works on Data and Communication Security) of IEC TC57 mainly for handling the security of TC 57 series of protocols including IEC 60870-5 and its derivatives,

IEC 60870-6 (TASE.2), and IEC 61850. In addition, security through network and system management has to be addressed. These security standards have been developed in such a way that it should meet different security objectives for the different protocols, which vary depending upon how they are used. Some of the security standards can be used across a few of the protocols, while others are very specific to a particular profile. The different security objectives include authentication of data transfer through digital signatures, ensuring only authenticated access, prevention of eavesdropping, prevention of playback and spoofing, and intrusion detection.

6.4.7 IEC 62056 ELECTRICITY METERING DATA EXCHANGE STANDARD

IEC 62056 is a set of standards for Electricity metering data exchange by International Electrotechnical Commission. The IEC 62056 standards are the International Standard versions of the DLMS/COSEM specification. DLMS or Device Language Message Specification (originally Distribution Line Message Specification), is the suite of standards developed and maintained by the DLMS User Association (DLMS UA) and has been adopted by the IEC TC13 WG14 into the IEC 62056 series of standards. The DLMS UA maintains a liaison with IEC TC13 WG14 responsible for international standards for meter data exchange and establishing the IEC 62056 series. In this role, the DLMS UA provides maintenance, registration and compliance certification services for IEC 62056 DLMS/COSEM.

COSEM or Companion Specification for Energy Metering includes a set of specifications that defines the Transport and Application Layers of the DLMS protocol. The DLMS UA defines the protocols into a set of four specification documents namely Green Book, Yellow Book, Blue Book and White Book. The Blue book describes the COSEM meter object model and the OBIS object identification system, the Green book describes the Architecture and Protocols, the Yellow book treats all the questions concerning conformance testing, the White book contains the glossary of terms. If a product passes the Conformance Test specified in the Yellow book, then a certification of DLMS/COSEM compliance is issued by the DLMS UA.

The IEC TC13 WG14 groups the DLMS specifications under the common heading: *Electricity metering data exchange – The DLMS/COSEM suite*. DLMS/COSEM protocol is not confined to electricity metering; rather it is used for gas, water and heat metering.

6.4.8   IEC 62056-21

IEC 61107 is the current IEC 62056-21, is an international standard for a computer protocol to read utility meters. It is designed to operate over any media, including the Internet. A meter sends ASCII or High level Data Link Control (HDLC) data to a nearby Hand-Held Unit (HHU) using a serial port. The physical media are usually either modulated light, sent with an LED and received with a photodiode, or a pair of wires, usually modulated by a 20mA current loop. The protocol is usually half-duplex.

6.5   SECURE COMMUNICATION (sCOMMUNICATION)

The main difference between DCS networks and IT networks is mainly the control part and which is not at all a surprise. As an example a compromised power system SCADA is an unacceptable threat to the reliability of the Bulk Electric System (BES). All software can be hacked, including firewalls, hence why best-practice and unconditional security standards are recommended. Obviously ICCP communication also meets all of these requirements. ICCP does not provide authentication or encryption. These services are normally provided by lower protocol layers. ICCP uses Bilateral Tables to control access. A Bilateral Table represents the agreement between two control centers connected with an ICCP link. The agreement identifies data elements and objects that can be accessed via the link and the level of access permitted. Once an ICCP link is established, the contents of the Bilateral Tables in the server and client provide complete control over what is accessible to each party. There must be matching entries in the server and client tables to provide access to data and objects.

6.6   SELECTING THE RIGHT PROTOCOL FOR SCADA

As there are so many protocols available as both proprietary and open, many factors are to be considered when choosing the protocol for SCADA. The following points are useful while designing the SCADA.

1. Determine the system area with which automation engineers are concerned, e.g.

    1. the protocol from a SCADA master control station to the SCADA RTUs,

    2. a protocol from substation IEDs to an RTU or a PLC, or a LAN in the substation.

2. As the technology is changing so fast that the timing of utility installation can have a great impact on the protocol which is selected. Hence the installation period must be determined most appropriately.

If new IEDs are implemented in the substation and scheduled to be in service within six months, Protocols may be selected accordingly as Modbus and Modbus Plus are suitable at present in Indian scenario. But if the period of installation and intended applications is for a long time, then consider IEC 61850 and UCA2 MMS as the protocol.

If the timeframe is around one year make protocol choices from implementing agency who acts as the industry initiatives and incorporates this technology into their product's migration paths. This helps to protect the investment from becoming obsolete by allowing incremental upgrades to new technologies.

In the design phase, protocol choices differ with the application areas. Different application areas are in different stages of development. An awareness of development stages will help to determine realistic plans and schedules for specific projects. Earlier when SCADA were designed, information and system security was not a priority. Most of the SCADA were designed as proprietary, stand-alone systems, and their security resulted from their physical and logical isolation and controlled access to them.

With the advancement of Information and Communication Technology (ICT), SCADA and DCS began to accept open standards and advanced networking technologies especially the Internet technology. Suppliers acquired the capability to implement Web-based applications to perform monitoring, control, and remote diagnostics. Obviously this introduces control system cyber vulnerabilities. In addition to traditional IT vulnerabilities, SCADA specific vulnerabilities become the real threat which triggers unique cyber security requirements to protect ICS against these SCADA specific attacks and vulnerabilities. Further it is a fact that todays technology may become obsolete tomorrow as the pace of the technological advancement is so fast, it seems the time between the present and the future is shrinking. Hence it is most important that one must evaluate not only the vendor's or implementation agencies present products but also their future product development and implementation strategies.

SUMMARY

This chapter begin with the evolution of communication protocols and then move on to explain the various communication protocols used today such as DNP3, Modbus, Profibus, IEC 60870, IEC 61850, and ICCP TASE.2 (IEC 60870-6). Other relevant ICS and PSS protocols such as IEEE C37.118.1 Synchrophasor Measurement Standard, IEC 61968 standard, IEC 61970 standard, IEC 62325 standard, IEC 61508, IEC 62351, IEC 62056 and IEC 62056-21 are also fleetingly presented. Finally the important points to select the right protocols are also described.

Chapter SEVEN

SECURING INDUSTRIAL CONTROL SYSTEMS

7.1 INTRODUCTION

Today it is a fact that SCADA networks have become an integral component in modern industrial environment to monitor and analyse real-time data, control both local and remote industrial processes, interact with various field devices, and keep logs and events for analysing, auditing and other purposes. For a safe and secure function of the SCADA network in an industrial environment, secure data transfer is a growing need of the time. The dawn of the extremist threat to critical infrastructures of many nations made SCADA systems no longer a hidden, unknown entities working secretly to control industrial and commercial operations. Further SCADA systems and Industrial Control networks are becoming the critical targets for malicious individuals, aggressive nations, terrorist groups, curious hackers, and organization's competitors.

This chapter explores the vulnerabilities and risks associated with SCADA systems, the changing mentality of SCADA network users, attack paths to SCADA components, and the potential actions that can be taken against plants employing SCADA technology. The chapter concludes with a discussion of a new application of honeypots and honeynets to protect SCADA systems and capture relevant information on attackers' approaches and methods that can be used to develop SCADA security controls that are more effective.

7.2   IT SECURITY AND SCADA SECURITY

The present warfront describes the cyberspace as the interdependent network of information technology infrastructure, which includes the Internet, telecommunications networks, computer systems, embedded processors and process/plant controllers of critical industries. Today this interdependent cyberspace is a crucial component of National Critical Infrastructure and needs distinct protective measures. Cyberspace is used to exchange information, buy and sell products and services, and enable many online transactions across a wide range of sectors, both nationally and internationally. This necessitates a secure cyberspace else, will indeed impact to the health of the economy and the security of any Nation. Hence to protect the Nation's critical information infrastructures, from risks such as online fraud, identity theft, and misuse of information online, a well-defined cyber security policy with an effective mitigation plan in accordance with the organisation's safety and security has to be implemented with utmost care. Today the rapid growth and amalgamation of ICT in SCADA and the social inter-dependency brought a paradigm shift to the perception of Critical Information Infrastructure threats. As a consequence, today cyber security is becoming a global program which is to be addressed collectively. These growing threats to security, at the level of the individual, the firms, government and critical infrastructures, essentially make security everyone's responsibility. Hence it is important to comprehend and keep familiar with the contours of fast changing challenges. Modern automation mainly depends on the SCADA technology and the security requirement of the SCADA is quite different from IT security as it involves mission critical processes or operations. The basic difference of the IT security and SCADA security are briefly described below.

Generally in SCADA systems, or ICS, the fact that any logic execution within the system has a direct impact in the physical world. This demands safe operations in the field or plant equipments. The field devices being on the frontline and directly affect the human lives and ecological environment, the field devices in SCADA systems are in no way inferior to the central hosts. Certain operating systems and applications running on SCADA systems, which are unconventional and proprietary to IT personnel, may not operate correctly with commercial off the-shelf IT cyber security solutions. Also, factors like the continuous availability demand, time-criticality, constrained computation resources on edge devices, large physical base, wide interface between digital and analog signals, social acceptance including cost effectiveness and user reluctance to change, legacy issues and so on make SCADA system very complex to make it secure. Host SCADA systems operates in real-time and completion of an operation after its deadline may be useless and potentially can cause cascading effect in the physical world. The operational deadlines from command to system response imposed with stringent constraints as a missing deadline can cause a complete failure of the system. Hence latency is extremely important and can be destructive to SCADA system's performance. If the system does not react within a certain time frame would cause great loss in safety, such as damaging the surroundings or threatening human lives. It's not the length of time frame but whether meeting the deadline or not distinguishes hard real-time system from soft real-time system. Soft real-time systems, may tolerate certain latency and respond with decreased service quality. Non-major violation of time constraints in soft real-time systems leads to degraded quality rather than system failure. Furthermore due to the physical nature, tasks performed by SCADA system and the processes within each task are often needed to be interrupted and restarted. The timing aspect and task interrupts can preclude the use of conventional encryption block algorithms. With Real-time operating system (RTOS), SCADA's vulnerability also rises from the fact that memory allocation is more critical in an RTOS than in other operating systems. Further many field devices used in SCADA today are embedded systems and run years without rebooting which accumulate fragmentation. Thus, buffer overflow is more problematic in SCADA than in traditional IT.

7.3   SECURITY DEFINITIONS

Automation engineers often use the cyber security terms vulnerability, threat, risk, etc. However very few use those terms in the right sense by comprehending the correct meaning and their relationships between them. Some of the important cyber security terms of importance are briefly explained below.

Vulnerability: It is a weakness in a SCADA system for a countermeasure and could be exploited by a threat. It can be a software, hardware, procedural, or human weakness that can be exploited.

Threat: It is any potential to destructively impact a SCADA system that is associated with the exploitation of a vulnerability. The threat is that someone, or something, will identify a specific vulnerability and use it against the company or individual. The entity that takes advantage of a vulnerability is referred to as a *threat agent*. A threat agent could be an intruder accessing the network through a port on the firewall, a process accessing data in a way that violates the security policy, or an employee making an accidental mistake that could expose confidential information.

Risk: It is the probability that a specific threat exploiting a vulnerability and the consequent impact or harm to the SCADA system. If a firewall has several ports open, there is a higher probability that an intruder will access the network in an unauthorized method.

Exposure: It is an instance of being exposed to losses. Vulnerability exposes an organization to possible damages. If password management is lax and password rules are not enforced, the company is exposed to the possibility of having users' passwords captured and used in an unauthorized manner. If a company does not have its wiring inspected and does not put proactive fire prevention steps into place, it exposes itself to potentially devastating fires.

Control or countermeasure: It is put into place to mitigate (reduce) the potential risk. A countermeasure may be a software configuration, a hardware device, or a procedure that eliminates vulnerability or that reduces the likelihood a threat agent will be able to exploit a vulnerability. Examples of countermeasures include strong password management, firewalls, a security guard, access control mechanisms, encryption, and security-awareness training.

Safeguard: A countermeasure or security control designed to reduce the risk associated with a specific threat.

Impact: The effect or consequence of a threat realized against a SCADA system.

If users are not aware of the processes and procedures, there are chances that an employee will make an unintentional mistake that may destroy data. If an Intrusion Detection System (IDS) is absent on a network, the attack probability is very high. Risk ties the vulnerability, threat, and likelihood of exploitation to the resulting business impact.

7.4   MANAGING RISK

The fundamental concepts of managing risk in the information security arena can be also applied efficiently to SCADA system security. The basic risk management concepts used to reduce risk in SCADA systems are fundamentally designed to reduce the adverse impact of threats that materialize and exploit the vulnerabilities inherent in SCADA systems. It involves assessing the risk, mitigating the risk, and then continuously evaluating system risks.

Assessing risk: Identifying the risks is vital for any businesses, but some risk are crucial to certain usiness and may be more vulnerable to some threats than others. Such risks must be identified, assessed and addressed, if not, resources and effort on threats will likely hit the business and may be disastrous.

Prioritising risk: While the threat landscape is constantly changing and posing new threats, it is not effective to simply look at the potential attacks that could hit from outside, rather it is vital to look inward also. Risk management puts a spotlight on where the biggest risk lies in the organisation and enables to protect it accordingly. Without this key step in risk management, businesses would be blindly defending across their valuable assets.

Mitigating risk: Human imperfection is often at the root of serious cyber-security problems, with forgetfulness and carelessness costing organisations valuable asset. Hence raising awareness within the organisation is the first step in mitigating risk on an extensive basis. This could include educating staff on phishing attacks, and keeping up to date with patching, etc. can prevent major cyber incidents to a considerable extent.

Align risk and strategy: A proper integration of risk management into the organisation's business strategy, allows for a flexible approach to protect the organisation by triggering the necessary processes. As a result, irrespective of changes in strategy and approaches, cyber security risk management never becomes detached and weak points are not exposed because the system is all-inclusive. A crucial reason for this flexibility is to remain formidable in the face of constantly more formidable threat landscape. The situation is no longer a two dimensional game of walls and shield around an enterprise, but the approach to security must be able to adapt to fit at all times.

Employee training: A successful risk management without an appropriate employee training is impossible. Since human nature can never be made cent percent reliable, exposing staff to knowledge of the correct procedures will also add to the overall mitigation of risks. With huge numbers of people entering the network of an organisation using a record number of different devices, it is imperative that staffs are being made responsible with what they are allowing to cross the digital threshold.

7.5   SCADA THREAT SOURCES

As explained earlier, a *threat* is any person, circumstance or event with the potential to cause loss or damage and a *vulnerability* is any weakness that can be exploited by an opponent. Both are evaluated based on the consequence and the amount of loss or damage occurred from a successful attack. Modern SCADA and ICS are typical Cyber-Physical Systems (CPS), which tightly integrates a physical process or plant with the cyber process of network computing and communication at all scales and levels. Obviously it is susceptible to cyber threats and vulnerabilities and can be exploited by different attack groups. They are mainly classified as follows.

1. *Threat from Crackers:* who break into computers for profit or bragging rights. Internet increases availability of hacker tools along with information about infrastructures and control systems,

2. *Ransomware threats:* Emergence of a strong financial motive for cybercrime to exploit vulnarabilities,

3. *Threat from Insiders:* who disrupt their corporate network, sometimes an accident, often for revenge,

4. *Threat from Terrorists:* who attack systems for cause or ideology, and

5. *Threat from Hostile Countries:* Attack computers of the enemy countries.

7.6   SCADA AND SMART GRID VULNERABILITIES, THREATS AND ATTACKS

With the integration of ICT, transition to a smarter electrical grid is very much optimistic and promising. The major points which lead to Smart Grid vulnerable to cyber-attacks are described briefly below.

1. ***Bidirectional communication:*** Though two way communications provides great benefits to the organization and the customer by giving the capability to communicate and share information, it makes the system vulnerable to cyber-attacks.

2. ***Customer data privacy:*** The information shared over the Smart Grid is intrinsically sensitive, requires high level privacy and personal security.

3. ***AMI:*** Various manufactures provides devices with different security features with different security levels built in, making it a challenge to standardize security practices.

4. ***Distributed connectivity:*** Smart meters are a part of the NAN in Smart Grid which is not confined to a specific geographic area. Hence the boundaries of the network will expand and become more difficult to secure.

5. ***Authentication and access controls:*** As the number of customers, suppliers and contractors increases, it becomes difficult to gain access to the network resulting in identity theft.

6. ***Proper employee training and awareness:*** Without proper training and awareness, there are chances of increasing the insider threats and lapses.

7. ***Guidelines, standards and interoperability:*** This may pose the potential for gaps in visibility, defense and recovery.

These threats are so real that many nations have declared its digital infrastructure as strategic asset and made cyber security a national security priority. They are setting up security policies to force power utilities responsible for protecting the critical electrical infrastructure. Many Nations entrusted their secret agencies to monitor for hackers attempting to infiltrate the power sector. But government regulation and pressure won't be enough. Hence utilities have to take spontaneous and significant steps to secure their networks. The techniques which attackers use to gain control of a SCADA or cause different levels of damage are mostly similar to those in case of IT. But they also possess certain explicit techniques and some of the techniques used are briefly described below.

1. *Spreading of Malwares:*   Malwares developed by an attacker can spread it to infect the AMI, or control center servers or utilities corporate servers. Malwares can replace or alter the device functions or a system including sending false and sensitive information.

2. *Backdoor entry through communication devices:*   A cyber-attacker may compromise some of the communication devices such as modems, routers, etc. and infiltrate the system using it as a backdoor to launch future attacks.

3. *Accessing and manipulating the database:*   All SCADA events and data are stored in a database on the control center server network and then mirror the logs into the business network. If the database management systems are not properly configured, a skillful cyber-attacker can gain access to the business network database, and then exploit the control system network.

4. *False Data Injection:*   An attacker can send false packets of information into the network, such as wrong smart meter data, false price tariffs, fake emergency event, etc. Fake information can have huge financial impact on the electricity markets. This type of attacking technologies is advancing much faster than security patch ups to control it. Hence it is very essential that the end node security aspects of Smart Grid especially the smart meter should ensure that it has communication capabilities that meet all basic integrity and confidentiality criteria. Unfortunately the smart meters today do not meet the required protection against false data injection. These facts highlight a much larger potential issues with data integrity throughout the smart grid infrastructure.

5. *DoS attacks Network Availability:*   DoS attacks might attempt to delay, block, or corrupt information transmission in order to make smart grid resources unavailable. As the smart grid uses IP protocol and TCP/IP stack, it becomes subject to all the vulnerabilities inherent in the TCP/IP stack and hence the DoS attacks.

6. *Modbus security issue:*   Modbus protocol is widely used in Power System SCADA especially for the communication between the IEDs and RTUs. Hence all Modbus security issues are applicable to the SCADA system or all the facility-based processes such as Smart Grid.

7. *Eavesdropping and traffic analysis:*   An enemy can obtain sensitive information by monitoring network traffic. It can be the utilities business strategy, future price information, control structure of the grid, and power usage. Later this data can be used for hostile deeds.

7.7   ALARMING SCADA AND SMART GRID THREATS

The present day Smart Grid threats are very advanced technically and the implemented Smart Grid and SCADA are secured just because of *security through obscurity*. A brief description of the technically advanced lethal threats and vulnerabilities are explained in the following sections.

7.7.1   ZERO DAY VULNERABILITIES

The term zero day implies the time before the developer could develop and deploy a patch up to overcome the flaw.

An attacker can and creates and deploy malwares by exploiting the flaw to attack the SCADA or ICS system. There are many zero-day flaws that may affect a SCADA system. Stack overflow is one of them. This attack can be launched on the field devices as well as the servers. The stack buffer in the memory can be corrupted by a malicious player, leading to injection of dangerous executable code into the running program and thus usurping the control of the industrial plant/process. Zero Day Attacks can also occur in the form of DoS attacks that overload computer resources.

Stuxnet is a lethal computer worm which uses four zero-day Windows vulnerabilities. It is believed that it had been primarily written to target Iranian nuclear centrifuges. Its final goal is to disrupt ICSs by modifying programs implemented especially on PLCs or RTUs to make them work in a manner that the attacker intended and to hide those changes from system operators. It is believed that Stuxnet is introduced to a computer network through an infected removable drive. To hide itself while spreading across the network and realizing the final target, the virus installs a Windows rootkit by exploiting four zero-day vulnerabilities. The success of this virus in penetrating the PLC environment shows that traditional security measures are not at all adequate for the ample protection and safety of the critical infrastructures.

7.7.2 NON-PRIORITIZATION OF TASKS

This is a serious flaw in real-time operating systems of many SCADA and ICS. In certain embedded operating systems, there may not be the feature of prioritization of tasks. Memory sharing between the equally privileged tasks lead to serious security issues. The features such as the accessibility to create Object Entry Point (OEP) in the kernel domain can lead to loopholes in security. Non-kernel tasks may be protected from overflows using guard pages. But the guard pages may be small and cannot provide stringent protection.

7.7.3 DATABASE INJECTION

Detrimental query statements can be injected to exploit the vulnerabilities in a SCADA especially when the client inputs are not properly filtered. This is widely reported for SQL-based databases. Here the attacker sends a command to SQL server through the web server and attempt to reveal critical authentication information.

7.7.4 COMMUNICATION PROTOCOL ISSUES

Today with the advancement in encryption and authentication, IT security is capable of encountering the sophisticated cyber-attacks and threats. But they are not adopted in an adequate manner in SCADA and Smart Grid especially when the process is controlled with the client server architecture. Earlier SCADA security was not a major concern and hence communication protocols were not given adequate importance to authentication. This does not mean that authentication and encryption methods cannot be used with these systems. It is very important to note that encryption is effective only in an authenticated communication between entities. To have a secure TCP/IP communication, Internet Protocol Security (IPsec) framework has to be employed. It will help to create a secure channel of communication for ICS. IPsec uses two protocols for authentication and encryption: Encapsulating Security Payload (ESP) and Authentication Header (AH). Advanced Persistent Threat (APT) attacks which monitor network activity and steal data for a future attack, can be effectively dealt with protocols like Syslog that keeps security logs which provide a means for detecting stealthy attempts to gather information prior to building sophisticated attacks by malicious players.

7.7.5 STEALTHY INTEGRITY ATTACKS

Stealth attacks are targeted to disrupt the service integrity, and make the networks to accept false data value. Security engineers have reported that very powerful antagonists equipped with thorough understanding, and disruption capabilities that are capable to perform stealthy attacks which partly or totally evade traditional abnormality detectors. In fact the severity of an attack strategy depends on the capabilities of enemies to coordinate attack vectors on control signals and sensor measurements.

7.7.6 REPLAY ATTACK

These are the network attacks in which an attacker spies the communication between the sender and receiver and takes the authenticated information. Typical example is stealing the key and then contacts the receiver with that key. In Replay attack, the attacker gives the proof of his identity and authenticity. The negative effect of a replay attack on a feedback control system has found that this attack strategy is carried out in two steps.

Step I The hacker records sensor measurements for a certain window of time before performing the attack.

Step II The hacker replaces actual sensor measurements with previously recorded signals while modifying controls signals to drive system states out of their normal values. A replay attack is capable of bypassing the classical detectors.

7.7.7 FALSE DATA INJECTION ATTACK

Distributed SCADA especially the PSS and Smart grid may often operate in hostile environments. The AMI components lacking tamper-resistance hardware increase the possibility to be compromised. With a compromised AMI, the attacker can inject false measurement reports to disrupt the SCADA operation. The objective of the attacker is to fool the state estimator by carefully injecting a certain amount of false data into sensor measurements. These attacks are denoted as false data injection attacks. It can disrupt the DCS system and leads to a false state estimation leading to upset the SCADA controller operations. Security engineers are of the opinion that the false data injection attack is a discrete-time state-space model driven by Gaussian noises.

7.7.8   Zero-Dynamics Attack

In SCADA Zero-Dynamics Attack (ZDA) is one of the toughest attacks to defend, especially in closed-loop feedback system which possesses an unstable zero, such as an unbounded actuator or sensor. The most unfortunate fact is that, it is quite unable to detect these attacks by scrutinising the data or data flow. An attacker having an in-depth knowledge of DAS can launch a zero dynamics attack in the cyber space or injected into the communication links. As modern control systems are implemented on digital computers using sample and hold mechanisms, where the controllers can be dealt with in a Sampled Data (SD) framework, they can generate vulnerability to stealthy attacks due to the unstable sampling zeros in the SD system.

7.7.9   Covert Attack

This attack is a targeted attack, but it is constructed and deployed using public tools. These are custom-made and minimally equipped. The strategy of Covert attack consists of coordinating control signals and sensor measurements into a concerted malicious attack. This attack is executed in two steps.

Step I   The state attack vector can be chosen freely based on malicious targets and available resources.

Step II   The sensor attack vector is designed in such a way that it can compensate for the effects of the state attack vector on the sensor measurements.

The covert attack strategy can be considered the worst-case attack because of its ability to bypass traditional anomaly detectors. However, the covert attack needs to compromise numerous sensors to assure its stealth. Therefore, SCADA defenders can remove a covert attack by protecting some critical sensors or deploying secure sensors.

7.7.10   Surge Attack, Bias Attack, and Geometric Attack

The three types of stealthy attacks, viz. the surge attack, the bias attack, and the geometric attack are also important and to be addressed by cyber-security engineers. The surge attack causes to maximize damage as soon as possible, while the bias attack slowly modifies the system by small perturbations over a long period. The geometric attack integrates the surge attack and the bias attack by shifting the system behaviour gradually at the beginning and optimizing the disaster at the end.

7.8 DREADFUL SCADA MALWARES

With the automation using SCADA the efficiency and performance of the power grid and industrial plants or process can be elevated to new operational heights, but they increase the system vulnerability to potential cyber-attacks. Black Energy, Stuxnet, Havex, Duqu, and Sandworm are all recent examples of malwares targeting SCADA systems. The AMI components especially the smart meters and increase in External Access Points (EAP) added with the integration of RES introduced new additional areas through which a potential cyber-attack may be launched on the grid. The present malware intrusions have resulted in a significant disruption of grid operations like what it had occurred to the nation Ukraine by the dreadful malware BlackEnergy.

7.9 PRIVILEGE TARGETS OF HACKERS

If a hacker could penetrate a SCADA or an ICS system successfully, the next step of the hacker would be to gain some degree of control of the SCADA system components. The extent of control attained is a function of the protections linked/embedded with each system component, its visibility to the attacker, and the capabilities and intentions of the attacker.

Various exploitations which usually accomplished by a hacker through a malicious attack on a SCADA system are enumerated below.

1. Obtain access to the SCADA system
2. Obtain access to SCADA master control station
3. Compromise the RTU or local PLCs
4. Compromise the SCADA master control station
5. Obtain SCADA system passwords from master control station
6. Obtain access to RTUs or local PLCs
7. Spoof RTU and send incorrect data to master control station
8. Spoof master control station and send incorrect data to RTU

9. Shut down the master control station

10. Shut down local control RTUs

11. Disrupt communications between SCADA master control station and RTUs and

12. Modify RTU control program

These actions and their relative harmful impact on a SCADA or DCS system are the Privilege Targets of Hackers.

7.9.1   ATTACK VECTOR THROUGH HMI

Today, within the various SCADA solutions, the HMI, especially the HMI at remote locations for local data monitoring and configuring is the most favourite and most preferred target for attackers. The HMI acts as a centralized hub for managing critical infrastructure. If an attacker succeeds in compromising the HMI, nearly anything can be done to the infrastructure itself, which causes physical damage to SCADA equipment. Even if attackers decide not to disrupt operations, they can still exploit the HMI to gather information about a system or disable alarms and notifications meant to alert operators of danger to SCADA equipment. Many researchers found that most HMI vulnerabilities fell into four categories and they are memory corruption, credential management, lack of authentication/authorization and insecure defaults, and code injection. All of which are preventable through secure development practices. It is also observed that the average time needed between disclosure of a Zero Day Initiative (ZDI) of a SCADA vendor and the time for releasing a patch takes up to 150 days. This delay in releasing patches must be minimized by the HMI vendors by giving special attention and respond accordingly. Also the ZDI researchers should start with basic fuzzing techniques to find new vulnerabilities in HMIs.

Better look, for new file associations during installation, to aid in fuzzing, as many of the file formats are wide open. Further HMI and SCADA solutions would be well advised to adopt the secure life cycle practices implemented by OS and application developers over the last decade. By taking simple steps such as auditing for the use of banned APIs, vendors can make their products more resilient to attacks.

SCADA engineers also need to expect their products to be used in manners that they did not intend.

Today the automation engineers, physical-cyber security professionals and SCADA vendors are working together to find and report bugs in HMI and develop patches so that the HMI security continue to improve. Though a completely secure system can never be achieved, with a robust research and development tactics, security engineers can keep the lights on as long as security is a must, else Human Machine Interface may become a Hacker Machine Interface.

7.9.2 SECURITY CONCERNS OF SCADA CONTROL CENTRE

The control centre architecture must be designed with utmost security and with proper disaster recovery centre. Secured ingress and egress must be ensured with the field devices, IEDs, Smart meters, other control centres, etc. Usually the field devices and smart meters communicate with the Communication Front End (CFE) server through VPN preferably SSL/VPN. Obviously the CFE should have the capability of handling very large data, and in certain situation act as a Data Management Server with Web server capabilities. Front End Processor of CFE server should have very high processing capability and must be hot redundant. Today it is a general practice that the control centre architecture is logically segmented to various zones and critical networks are isolated. The data transfer between these zones is generally through firewalls with proper configuration. One of the usual recommended practice is to separate the SCADA network from the corporate network.

The nature of network traffic on these two networks should be different. Internet access, FTP, e-mail, and remote access will typically be permitted on the corporate network but should not be on the SCADA network. Rigorous change control procedures for network equipment, configuration, and software changes may not be in place on the corporate network. If SCADA network traffic is carried on the corporate network, it could be intercepted or be subjected to a denial of service attack. By having separate networks, security and performance problems on the corporate network would not be able to affect the SCADA network. A typical control centre architecture of power system SCADA having *Firewall with DMZ between Corporate Network and Control Network* are shown in Figure 7.1.

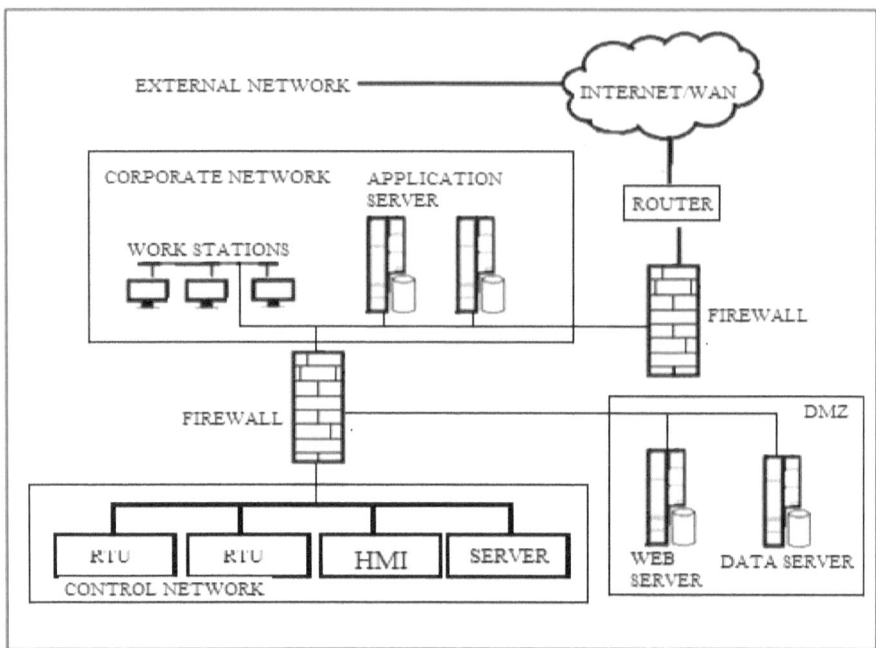

Figure 7.1   Firewall with DMZ between Corporate and Control Networks

The use of firewalls with the ability to establish a DMZ between the corporate and control networks. Each DMZ holds one or more critical components, such as the data historian, the wireless access point, or remote and third party access systems. In effect, the use of a DMZ-capable firewall allows the creation of an intermediate network.

Creating a DMZ requires that the firewall to offer three or more interfaces, rather than the typical public and private interfaces. One of the interfaces is connected to the corporate network, the second to the control network, and the remaining interfaces to the shared or insecure devices such as the data historian server or wireless access points on the DMZ network.

Non-firewall-based solutions will generally not provide suitable isolation between control networks and corporate networks. The two-zone solutions (no DMZ) are marginally acceptable but should be only deployed deployed with extreme care. The most secure, manageable, and scalable control network and corporate network segregation architectures are typically based on a system with at least three zones, incorporating a DMZ.

7.9.3 FLASH DRIVE USAGE AND END NODE SECURITY (ENS)

USB attacks are becoming more sophisticated, affecting all classes of USB device instead of just storage. As USBs have become a common method for easily sharing information locally between devices, they have become a common source of information system cyber compromise. As per NERC CIP guidelines, use of USB or USB type ports are strongly discouraged because a USB port is not immune to protection from *unauthorized access*. It would be helpless against connecting modems, network cables that bridge networks or insertion of an infected USB pen drive. Cyber protection for USB ports can be enforced, however, it is often cost prohibitive and is not one hundred percent effective. There is no essential requirement for using a USB instead of other standard and more secure interfaces such as Ethernet and serial ports. Some of the common methods to protect the USB ports are:

1. Disabling (via software) the physical ports

2. Prominent physical port usage discouragement such as a port cover plate or tamper tape

3. Physical port obstruction using removable locks

These measures are examples of defense-in-depth methods, but the CIP guidelines acknowledge that these control approaches can be easily circumvented. It is also not uncommon for an employee or authorized contractor to inadvertently compromise a device simply by plugging in an infected smart phone to charge the battery. USB flash drives pose two major challenges to critical infrastructure cyber security:

1. ease of data theft owing to their small size and transportability, and

2. system compromise through infections from computer viruses, malware and spyware.

It is a well-accepted fact that a USB supported portable peripheral device can trigger a massive cyber-attack, even when the computer system targeted is isolated and protected from the outside with firewalls and other types of security devices.

7.9.4 BADUSB

BadUSB is an USB which includes firmware in addition to disk space. It is inherently a microcontroller with writable storage memory registers.

This firmware however can be embedded with executable codes which cannot be verified by third party security software applications since the firmware is not open source. This flaw in USBs opens the door to modification USB firmware, which can easily be done from inside the operating system, and hide the malware in a way that it becomes almost impossible to detect. The flaw is even more potent because complete formatting or deleting the content of a USB device won't eliminate the malicious code, since it's embedded in the firmware. Patches made for BadUSB have been largely ineffective and a fix is years away. In fact till date, there has been no practical defensive solution against BadUSB attacks and it exposes the fundamental vulnerabilities of unconstrained privileges in USB devices. Under this the situation, PSS and Smart Grid design and implementation must be in such a manner that it completely eliminate the need for a USB port is advised and recommended in the interest of reliable and safe operations.

7.10   SCADA INTRUSION DETECTION SYSTEMS (IDS)

An intrusion detection system (IDS) is typically an appropriate combination of software and hardware or software alone which is capable to detect, analyse, keep log, and report the malicious events occurring in a network or in a host computer. IDS are also capable of scanning to detect open ports or other means of entry into the network or computer.

The logs of attack or attempts to attack help the security engineers to device strategies to prevent attacks. The details of the log is usually communicated to the security engineers or stored in a secured location for suitable analysis.

IDS are mainly classified into two, depending upon the location where they are deployed and source from where data is received for analysis. The IDS which are deployed in the network to monitor the network traffic is refereed as Network Intrusion Detection System (NIDS). If deployed in the host computer, then it is refereed as Host Intrusion Detection System (HIDS). Another classification is based on the strategy of detecting the malicious threats. The two various technologies generally used by the IDS are signature based and anomaly based. IDS can also be classified as passive, active and hybrid.

An NIDS captures data packets travelling over a network segment and evaluates, are mainly passive devices that use sensors to monitor network traffic. They are mainly implemented for protecting the host computers and servers from directly hitting with malicious worms and virus. One of the major disadvantages is that during the high network traffic, NIDS may not be able to monitor all the data packets. This miss can cause an attack. Further an NIDS are not capable of monitoring the encrypted packets. As a result it is not capable of preventing the attacks with the encrypted messages. NIDS are also incapable of predicting the impacts, though it can inform the security engineers whether the attack has been launched. On the other hand HIPS which are residing the host computers or servers are capable of monitoring and analysing the data more efficiently and give details such as entities affected, the outcome of the attack, etc. Further HIDS are capable of handling the encrypted data packets. HIDS utilises the information from the logs and operating system audit trials. The major disadvantage of HIDS is that it takes considerable time for the monitoring and evaluating the risks and outcome. Hence inorder to meet the latency requirements, host computers which are loaded with HIDS should have high computing speed and capabilities. Obviously it adds to the cost of the host computers and servers.

Signature-based IDS: This is a database where all known attack patterns are stored for comparing with the activity characteristics or the patterns which is detected by IDS. If any correlation is found, then it provides the specific and reliable information to the security engineers for necessary action or executes appropriate action if the IDS is an active one. But the major limitation is the inability of detecting the new attacks as its signature is not there in the database.

Anomaly-based IDS: It carefully monitors the network and host data flow and takes statistical samples to develop a profile of normal activities on a system. If these network or host statistics deviate from the custom, it usually indicates that an attack is in progress. Statistics used to characterize normal behaviour include CPU utilization rate and the number of failed login attempts. The major advantage of these IDS is that it can detect new attacks, but it can cunningly be deceived with attacks which do not significantly change the parameters being measured. Further, an anomaly-based IDS generates many false alarms, as certain genuine activity causes the statistical parameters to change enough to trigger an attack alarm.

Active-Response IDS: It takes some form of action in the event of an attack. The typical responses are briefed below.

Explore the situation and gather maximum information that would be pertinent in identifying the attack.

Block network ports and protocols used by the suspected attacker.

Change router and firewall access lists to block messages from the doubted attacker's IPaddress.

Passive–Response IDS: It provides information about an attack to the security engineer concerned who can then decide the proper course of action. The security engineer can be notified in the form of a call, an e-mail message, an alert on a computer terminal screen, or a message to a simple network management protocol (SNMP) console. The alert may include the source IPaddress of the attack, the IPaddress of the target of the attack and the result of the attack.

Analysing the IDS information in block form after an attack occur in batch-mode or interval-based processing. In this case, the user cannot intervene during an event as it is not a real-time operation. Batch-mode was common in early IDS implementations because their capabilities did not support real-time data acquisition and analysis.

7.11   DEFENCE-IN-DEPTH ARCHITECTURE

Another SCADA control centre architecture which gained much popularity with the incorporation of modern firewall technology is the defence in depth architecture. A single security product, technology or solution cannot adequately protect a SCADA by itself. A multiple layer strategy involving two (or more) different overlapping security mechanisms, a technique also known as defence-in-depth, is desired so that the impact of a failure in any one mechanism is minimized. A defence-in-depth architecture strategy includes the use of firewalls, the creation demilitarized zones, intrusion detection capabilities along with effective security policies, training programs and incident response mechanisms. In addition, an effective defence-in-depth strategy requires a thorough understanding of possible attack vectors on a SCADA. These include:

1. Backdoors and holes in network perimeter,
2. Vulnerabilities in common protocols,
3. Attacks on Field Devices,

4. Database Attacks, and

5. Communications hijacking and 'Man-in-the-middle' attacks.

Figure 7.2 shows a SCADA defence-in-depth architecture strategy that has been developed with improved Control *Systems Cyber Security.* The *Control Systems Cyber Security using Defence in Depth Strategies* for organizations use control system networks which maintain a multi-tier information architecture that requires:

1. Maintenance of various field devices, telemetry collection, and/or industrial-level process systems,

2. Access to facilities via remote data link or modem, and

3. Public facing services for customer or corporate operations.

This strategy includes firewalls, the use of demilitarized zones and intrusion detection capabilities throughout the ICS architecture. The use of several demilitarized zones in Figure 7.2 provides the added capability to separate functionalities and access privileges and has proved to be very effective in protecting large architectures comprised for networks with different operational mandates. Intrusion detection deployments apply different rule-sets and signatures unique to each domain being monitored.

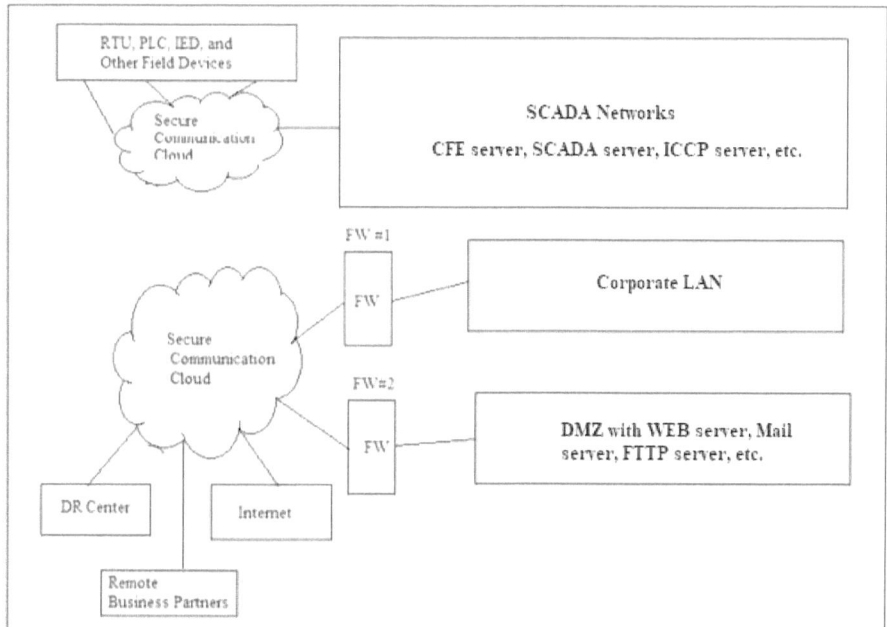

Figure 7.2 Defense-in-Depth Architecture

7.12   FIREWALL DEPLOYMENT AND FIREWALL POLICIES

Once the defence-in-depth architecture is in place, the work of determining exactly what traffic should be allowed through the firewalls begins, configuring the firewalls to deny all except for the traffic absolutely required for business needs of every organization. Security engineers of the organisation must be aware of the data flow requirements of business and the security impacts of allowing that traffic through. Typical example is, many organizations considered allowing SQL traffic through the firewall as required for business for many data historian servers. Unfortunately, SQL was also the vector for the Slammer worm. Many important protocols used in the industrial world, such as HTTP, FTP, OPC/DCOM, Ethernet/IP, and MODBUS/TCP, have significant security vulnerabilities.

The following sections summarize some of the key points on Firewall Deployment for SCADA and Process Control Networks. When installing a single two-port firewall without a DMZ for shared servers particular care needs to be taken with the rule design. At a minimum, all rules should be stateful rules that are both IP address and port (application) specific. The address portion of the rules should restrict incoming traffic to a very small set of shared devices (e.g., the data historian) on the control network from a controlled set of addresses on the corporate network. Allowing any IP addresses on the corporate network to access servers inside the control network is not recommended. In addition, the allowed ports should be carefully restricted to relatively secure protocols such as Hypertext Transfer Protocol Secure (HTTPS). Allowing HTTP, FTP, or any unencrypted SCADA protocol to cross the firewall is a security risk due to the potential for traffic sniffing and modification. Rules should be added to deny inbound communication with the control network. Rules should allow internal devices in the control network to establish connections outside the control network in a most secured manner only. On the other hand, if the DMZ architecture is being used, then it is possible to configure the system so that no traffic will go directly between the corporate network and the control network. With a few special exceptions which are mentioned below, all traffic from either side can terminate at the servers in the DMZ. This allows more flexibility in the protocols allowed through the firewall. For example, MODBUS/TCP might be used to communicate from the RTUs/IEDs to the data historian, while HTTP might be used for communication between the historian and enterprise clients. Both protocols are inherently insecure, yet in this case they can be used safely because neither actually crosses between the two networks. An extension to this concept is the idea of using *disjoint* protocols in all control networks to corporate network communications. That is, if a protocol is allowed between the control network and DMZ, then it is explicitly **not** allowed between the DMZ and corporate network. This design greatly reduces the chance of a worm such as Slammer actually making its way into the control network, since the worm would have to use two different exploits over two different protocols.

One area of considerable variation in practice is the control of outbound traffic from the control network, which could represent a significant risk if unmanaged. One example is Trojan horse software that uses HTTP tunnelling to exploit poorly defined outbound rules. Thus, it is important that outbound rules be as stringent as inbound rules. Technical Report contains some example guidelines that help clarify this. A summary of the preferred outbound and inbound rules for the secure data flow are described below.

1. Inbound traffic to the control system should be blocked. Access to devices inside the control system should be through a DMZ,

2. Outbound traffic through the control network firewall should be limited to essential communications only, and

3. All outbound traffic from the control network to the corporate network should be source and destination-restricted by service and port.

In addition to these rules, the firewall should be configured with outbound filtering to stop forged IP packets from leaving the control network or the DMZ. In practice this is achieved by checking the source IP addresses of outgoing packets against the firewall's respective network interface address. The intent is to prevent the control network from being the source of spoofed communications, which are often used in DoS attacks. Thus, the firewalls should be configured to forward IP packets only if those packets have a correct source IP address for the control network or DMZ networks. Finally, Internet access by devices on the control network should be strongly discouraged. Usually the firewalls deployed in PSS come with the default configuration as described below. However the installation and security engineers do confirm the ruleset of the firewalls without any compromise, else can be catastrophic.

1. The base rule set should be deny all, permit none.

2. Ports and services between the control network environment and the corporate network should be enabled and permissions granted on a specific case-by-case basis. There should be a documented business justification with risk analysis and a responsible person for each permitted incoming or outgoing data flow.

3. All *permit* rules should be both IP address and TCP/UDP port specific, and stateful if appropriate.

4. All rules should restrict traffic to a specific IP address or range of addresses.

5. Traffic should be prevented from transiting directly from the control network to the corporate network. All traffic should terminate in the DMZ.

6. Any protocol allowed between the control network and DMZ should explicitly NOT be allowed between the DMZ and corporate networks (and vice-versa).

7. All outbound traffic from the control network to the corporate network should be source and destination-restricted by service and port.

8. Outbound packets from the control network or DMZ should be allowed only if those packets have a correct source IP address that is assigned to the control network or DMZ devices.

9. Control network devices should not be allowed to access the Internet.

10. Control networks should not be directly connected to the Internet, even if protected via a firewall.

11. All firewall management traffic should be carried on either a separate, secured management network or over an encrypted network with two-factor authentication. Traffic should also be restricted by IP address to specific management stations. These are only guidelines; hence a vigilant assessment of each control environment is required before finalizing and implementing the firewall ruleset.

7.13 PROPOSED SECURITY SOLUTIONS

Security solutions developed for traditional IT networks are not effective in Smart Grid and PSS because of the major differences between them. Their security objectives are different in the sense that, security in IT networks aims to enforce the three security principles viz. confidentiality, integrity and availability, while the security in PSS and Smart Grid networks aims to provide human safety, equipment and power lines protection, system operation, etc. Furthermore, the security architecture of IT networks is different than that of the Grid network since security in IT networks is achieved by providing more protection at the centre of the network where the data is kept, while the protection in automation networks is done both at the network centre and end nodes or edge. Their underlying topology is also different as IT networks use a well-defined set of operating systems and protocols, while Smart Grid networks use multiple propriety operating systems and protocol specific to vendors. Hence Quality of Service (QoS) metrics are different in the sense that it is acceptable in IT networks to reboot devices in case of failure or upgrade, while this is not at all acceptable in PSS and Smart Grid networks since services must be available 24×7. These major differences between the IT and grid network security objectives necessitate the need for new security solutions specific for the Smart Grid network. The development of Smart Grid security solutions faces many challenges and they are:

1. Some components use propriety operating systems to control functionality having no security features,

2. Most of the legacy PSS network was designed without regard to security,

3. Security should be integrated with existing systems without downgrading the latency and efficiency and performance,

4. Remote access to grid devices must be monitored and controlled, and

5. The new protocols selected should have the capability of incorporating future security solution.

7.13.1   SECURING THE AMI

Implementing a smart grid without proper AMI security could result in grid instability, loss of private information, utility fraud, and unauthorized access to energy consumption data. Without the proper security, the benefits of the trusted by-directional communication between consumers and utility, including the secure collection of information for accurate big data analytics, cannot materialize. As already explained AMI facilitates DR management by collecting data from smart meters across the entire grid. Hence the absence of AMI security can cause potential damage against infrastructures and privacy of Smart Grid by the adversaries.

Integrating a proper Role Based Access Control (RBAC) with the MDM system is one of the means of ensuring the smart meters' data. In other way Data security can be achieved through a proper assignment of roles and permissions so that meters' data is accessed only by authorized users. Smart meter being the critical device, the following threat vectors pertaining to the smart meters are to be addressed with utmost importance.

1. Usually all smart meters have an optical communication port with Ethernet connectivity to the head end system. The attackers may break the password, reconfigure the smart meter software and use it for stealthy exploit.

2. As the physical location of the smart meter need not be always secured, physical attacks can be launched by the attackers and read out the firmware, the system configuration, and the system credentials. These data also can be used by the attackers for gaining control over the remote access.

3. The head end connectivity of the smart meter is usually fiber optic link through VPN. If the data encryption is weak, the attackers may break and decrypt the data within the communication network. This attack provides the potential of compromising other devices in the communication network.

Mitigation techniques of AMI include providing a strong encryption, measures to prevent public disclosure of smart meter vulnerabilities, providing utmost attention to keep the credentials and key materials secret. Further the head end systems must be kept secret physically and from malware infection.

Internal attackers like disgruntled and rogue employees may send wrong commands to disconnect over the communication network to smart meters can cause a power cut due to the remote controllable breakers within the smart meter. A close monitoring and proper training is hence most essential to avoid such situations.

7.13.2 MAKING THE SMART GRID SMARTER THAN CYBER-ATTACKS

Ensuring the Smart Grid security can be achieved, by making the Smart Grid security smarter than the Smart Grid cyber-attacks. To achieve this following specific actions are suggested by cyber security experts and generally adopted. Certainly only a brave hearted person can implement these challenging requirements.

1. Conduct a thorough risk analysis to assess the risk and necessity of each connection to the SCADA network and develop a comprehensive understanding of all connections to the SCADA network, and how well these connections are protected especially to identify and evaluate the following types of connections.

   1. Internal local area and wide area networks, including business networks,

   2. The Internet,

   3. Wireless network devices, including satellite uplinks,

   4. Modem or dial-up connections, and

   5. Connections to business partners, vendors or regulatory agencies.

2. Strictly isolate the PSS network from other network connections to the maximum extent possible. Any connection to another network introduces security risks as it can add an EAP. Further if the connection creates a pathway from or to the Internet, it opens a channel to the public network which requires stringent firewalling with an uncompromising firewall ruleset. This requires a very high level of expertise. In addition, direct connections with other networks may allow important information to be lost. Hence isolation of the PSS network must be a primary goal to have a fool proof cyber protection. Strategies such as utilization of *demilitarized zones*

(DMZs) and data warehousing can facilitate the secure transfer of data between networks. However, DMZs must be designed and implemented very appropriately to avoid introduction of additional risk through improper configuration.

3. Carry out penetration testing or vulnerability analysis to all suspicious and external connections to the Smart Grid to evaluate the protection levels associated with these pathways. Use these information in conjunction with risk management processes to develop a robust security policy and protection strategy for any pathways to the Smart Grid or PSS networks. Since the PSS network is only as secure as its weakest connecting point, it is essential to implement firewalls, intrusion detection systems (IDSs), and other appropriate security measures at each point of entry especially at the end nodes and EAPs. Configure firewalls with appropriate ruleset to strictly prohibit access from and to the PSS network. Strategically place IDSs without compromising the efficiency especially the latency at each entry point to alert security personnel of potential breaches of network security.

4. PSS control servers built on open-source operating systems can be exposed to attack through default network services. To the greatest degree possible, remove or disable unused services to reduce the risk of direct attack especially when PSS networks are interconnected with other networks. Do not permit a service or feature on a PSS network without a thorough risk assessment of the consequences. Additionally, work closely with PSS vendors to identify secure configurations and coordinate any and all changes to operational systems to ensure that removing or disabling services does not cause downtime, interruption of service, or loss of support.

5. Some PSS use unique, proprietary protocols for communications between field devices and servers. Often the security of PSS is based solely on the secrecy of these protocols. Unfortunately, obscure protocols provide very little *real* security. Do not rely on proprietary protocols or factory default configuration settings to protect the system. Additionally, demand that vendors disclose any backdoors or vendor interfaces to the PSS, and expect them to provide systems that are capable of being secured.

6. Legacy PSS have no security features whatsoever. PSS owners must insist that their system vendor implement security features in the form of product patches or upgrades. Some newer PSS devices are shipped with basic security features, but these are usually disabled to ensure ease of installation. Analyze each PSS device to determine whether security features are present. Additionally, factory default security settings such as in computer network firewalls are often set to provide maximum usability, but minimal security. Set all security features to provide the maximum level of security. Allow settings below maximum security only after a thorough risk assessment of the consequences of reducing the security level.

7. Where backdoors or vendor connections do exist in PSS systems, strong authentication must be implemented to ensure secure communications. Modems, wireless, and wired networks used for communications and maintenance represent a significant vulnerability to the SCADA network and remote sites. Successful *war dialing* or *war driving* attacks could allow an attacker to bypass all other controls and have direct access to the PSS network or resources. To minimize the risk of such attacks, disable inbound access and replace it with some type of call-back system.

8. To be able to effectively respond to cyber-attacks, establish an intrusion detection strategy that includes alerting network administrators of malicious network activity originating from internal or external sources. A 24X7 intrusion detection system monitoring is essential incident response procedures must be implemented to allow an effective response to any attack. To complement network monitoring, enable logging on all systems and audit system logs daily to detect suspicious activity as soon as possible.

9. Technical audits of PSS devices and networks are critical to on-going security effectiveness. Many commercial and open-source security tools are available that allow system administrators to conduct audits of their systems/networks to identify active services, patch level, and common vulnerabilities. The use of these tools will not solve systemic problems, but will eliminate the *paths of least resistance* that an attacker could exploit. Analyze identified

vulnerabilities to determine their significance, and take corrective actions as appropriate. Track corrective actions and analyze this information to identify trends additionally, retest systems after corrective actions have been taken to ensure that vulnerabilities were actually eliminated. Scan non-production environments actively to identify and address potential problems.

10. Any location that has a connection to the PSS or Smart Grid network is a target, especially unmanned or unguarded remote sites. Conduct a physical security survey and identify all EAPs at each facility that has a connection to the PSS. Identify and assess any source of information including remote communication network connectivity that could be tapped. Identify and eliminate all single points of failure. The security of the site must be adequate to detect or prevent unauthorized access.

11. A fundamental management process needed to maintain a secure network is configuration management. Configuration management needs to cover both hardware configurations and software configurations. Changes to hardware or software can easily introduce vulnerabilities that undermine network security. Processes are required to evaluate and control any change to ensure that the network remains secure. Configuration management begins with well-tested and documented security baselines for the various systems.

12. Establish a disaster recovery plan that allows for rapid recovery from any emergency (including a cyber-attack).System backups are an essential part of any plan and allow rapid reconstruction of the network. Routine exercise disaster recovery plans may be carried out to ensure that they work and that personnel are familiar with them. Make appropriate changes to disaster recovery plans based on lessons learned from exercises.

13. Adopt the strategy of defence-in-depth. Defence-in depth must be considered early in the design phase of the development process, and must be an integral consideration in all technical decision-making associated with the network. Single points of failure must be avoided, and cyber security defence must be layered to limit and contain the impact of any security incident.

7.13.3   ZONE BASED ARCHITECTURE

Employing zone based architecture is one of the means of managing different parts of the grid, and provides protection. These zones can be categorized as customer operations, business, communication and control systems. Each zone is independently comprised of field devices, systems, communication media, and data centres which serve the specific operational functions of that zone. Implementing a set of security features common to all zones and security features required explicit to that zone can form an effective, modular and segmented security system. The following modular security zones are minimum suggested.

1. Customer operations security zone,
2. Corporate security zone, and
3. Communication and control systems security zone

Customer operations security zone: The customer operations security zone contains devices and processes that extend energy management to customers. It defines all functionalities of AMI. To provide security, access to home systems and the data gathered by AMI must be limited to authorized people and devices. Customer energy management systems must ensure integrity of command and meter data, authenticate devices, and protect the grid from compromised devices.

Corporate security zone: The corporate security zone includes all of the features and functions of the customer operations security zone plus security for IT functions vital to a business, such as email, internet, telephony, messaging, and a wide variety of corporate applications. To meet these requirements, consistent security policies must be applied across the entire product line in the smart grid.

Communication and control systems security zone: The communication and control systems security zone defines the processes used to manage the routing of energy from the generation plant to the consumer. It contains data centres involved in the generation, transmission, and distribution of energy, along with intelligent end devices to control energy flow and ensure grid reliability. Information collected and processed here supports equipment maintenance, troubleshooting, load capacity, and power re-routing in the event of outages.

To protect the integrity of telemetry data and control, utilities must ensure individual and device authentication, computer health verification, correlation of alarm data with other sensors to prevent false positives, regulatory compliance and enable forensic analysis. Data encryption, intrusion detection and prevention, and secure sharing of information between various data centres are critical.

As utilities make the transformation to smart grids, they need a foundation of converged IP networks, proven security principles, industry-leading networking equipment and software with integrated security capabilities to build end-to-end, secure grids. The maturity, reliability, and success of these products and services can shorten the learning curve for grid operators and allow energy companies to evolve operations to meet new standards and regulations.

7.13.4 FOLLOW STANDARDS AND GUIDELINES

Strictly follow security standards and Guidelines to ensure the Smart Grid security requirements without any compromise. It will indeed help to make Smart Grid smarter against cyber-attacks. The chief cyber security expert of the utility must be very keen in enforcing the following steps without compromise while designing and implementing the DCS.

1. Develop a Security Policy,
2. Establish Physical Security,
3. Lockdown Perimeter Security,
4. Enable Existing Security Features,
5. Secure Operational Traffic,
6. Secure Management Traffic,
7. Manage the CS Configuration,
8. Eliminate Security Shortfalls.
9. Continuous Security Training, and
10. Perform Security Audits.

Moreover to build a secure DCS strictly follow the key reminders without any compromise as listed and shown in the Figure 7.3.

1. Be always on the Defensive,
2. Be always vigilant for cyber-attacks,
3. Have safety, security and disaster recovery plans,
4. Implement physical-cyber security without any compromise,
5. Proper documentation and reporting of cyber incidents, attempts and attacks.

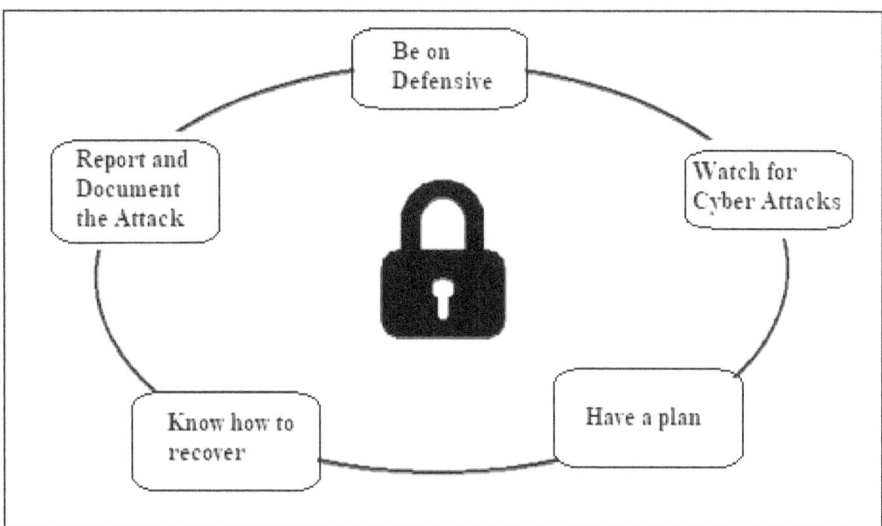

Figure 7.3 Smart Grid Security Key Reminders

SUMMARY

This chapter gives a description about how the SCADA security is different from IT security and its importance. The requirement of open communication system and standardization are discussed with emphasis on security. Security concerns of the substation automation and control centre architecture are discussed with solutions. Malware threats especially the attack of the lethal malware Stuxnet is a nightmare for SCADA implementing agencies because of the ZDVs of the Windows OS. These attack vectors and proposed solutions are discussed including threats and vulnerabilities of ICS and power system SCADA with various types of attacks and mitigating techniques are also discussed. Various proposals of the security engineers for making the automated power system smarter than cyber-attacks also have been elaborated.

INDEX

www.ingramcontent.com/pod-product-compliance
Lightning Source LLC
Chambersburg PA
CBHW021357210526
45463CB00001B/128